The Electricity at Work Regulations 1989

Guidance on Regulations

London: TSO

Published by TSO (The Stationery Office), part of Williams Lea, and available from:

Online
www.tsoshop.co.uk

Mail, Telephone, Fax & E-mail
TSO
PO Box 29, Norwich, NR3 1GN
Telephone orders/General enquiries: 0333 202 5070
E-mail: customer.services@tso.co.uk
Textphone 0333 202 5077

TSO@Blackwell and other Accredited Agents

© Crown copyright 2015

First published 1989
Third edition 2015

ISBN 978 0 7176 6636 2

This information is licensed under the Open Government Licence v3.0. To view this licence, visit http://www.nationalarchives.gov.uk/doc/open-government-licence/ **OGL**

Any enquiries regarding this publication should be sent to: copyright@hse.gov.uk

Some images and illustrations in this publication may not be owned by the Crown and cannot be reproduced without permission of the copyright owner. Where we have identified any third party copyright information you will need to obtain permission from the copyright holders concerned. Enquiries should be sent to copyright@hse.gov.uk

Printed in the United Kingdom for The Stationery Office.
SD000289 6/25

Guidance
This guidance is issued by the Health and Safety Executive. Following the guidance is not compulsory, unless specifically stated, and you are free to take other action. But if you do follow the guidance you will normally be doing enough to comply with the law. Health and safety inspectors seek to secure compliance with the law and may refer to this guidance.

Contents

Introduction 5

The Regulations 7

Regulation 1   Citation and commencement   7

Regulation 2   Interpretation   7

Regulation 3   Persons on whom duties are imposed by these Regulations   13

Regulation 4   Systems, work activities and protective equipment   15

Regulation 5   Strength and capability of electrical equipment   18

Regulation 6   Adverse or hazardous environments   19

Regulation 7   Insulation, protection and placing of conductors   22

Regulation 8   Earthing or other suitable precautions   25

Regulation 9   Integrity of referenced conductors   29

Regulation 10   Connections   30

Regulation 11   Means for protecting from excess of current   31

Regulation 12   Means for cutting off the supply and for isolation   33

Regulation 13   Precautions for work on equipment made dead   35

Regulation 14   Work on or near live conductors   37

Regulation 15   Working space, access and lighting   41

Regulation 16   Persons to be competent to prevent danger and injury   42

Regulations 17-28   43

Regulation 29   Defence   43

Regulation 30   Exemption certificates   44

Regulation 31   Extension outside Great Britain   44

Regulation 32   Disapplication of duties   45

Regulation 33   Revocations and modifications   45

Appendix 1   46

References and further reading   48

Further information   52

Introduction

About this book

1 This book gives guidance on and sets out the Electricity at Work Regulations 1989. The guidance is relevant to all work activities and premises except certain offshore workplaces and certain ships. Additional guidance specific to mines is given in *Electrical safety in mines*.[1]

2 The Electricity at Work Regulations 1989 (SI 1989/635) (as amended) (the Regulations) came into force on 1 April 1990. The purpose of the Regulations is to require precautions to be taken against the risk of death or personal injury from electricity in work activities. The text of the Regulations is available free to download from legislation.gov.uk or to purchase from The Stationery Office.

3 The Regulations are made under the Health and Safety at Work etc Act 1974 (the HSW Act).[2] The HSW Act applies principally to employers, the self-employed and to employees, including certain classes of trainees. Duties are imposed on people (referred to in this guidance as 'dutyholders') in respect of **systems**, **electrical equipment** and **conductors**, and in respect of work activities on or near electrical equipment. Words in bold (above and in extracts from the Regulations) are defined in regulation 2. The duties are in addition to those imposed by the HSW Act. The 1989 Regulations pre-date the risk assessment process brought in with the 1992 version of the Management of Health and Safety at Work Regulations. However, the risk assessment process is still required. Further information can be found in *Managing for health and safety*.[3]

What are the differences between this edition and the previous edition?

4 This third edition notes that regulations 17–28 and Schedule 1, which applied only to mines, were revoked in April 2015 by the Mines Regulations 2014. These Regulations also modified regulations 3 and 29.

Who should read this book?

5 The guidance is primarily for dutyholders (including those involved in the design, construction, operation or maintenance of electrical systems and equipment), engineers, technicians and their managers. It sets out the Regulations and gives technical and legal guidance on them. While it reflects the Health and Safety Executive's (HSE's) view of the meaning of terms used in the Regulations, only the Courts can provide a binding interpretation. The text of the Regulations is set out in *italics*, the accompanying guidance is in normal type. Coloured borders also indicate each section clearly.

Scope

6   This guidance applies to all operational voltages.

7   The Regulations apply to all **electrical systems and equipment** (as defined) whenever manufactured, purchased, installed or taken into use even if its manufacture or installation pre-dates the Regulations. Where electrical equipment pre-dates the Regulations this does not of itself mean that the continued use of the equipment would be in contravention of the Regulations. For example, some of the equipment to which the Regulations apply may have been made to a standard, such as a British Standard, which has since been modified or superseded. Standards such as BS 7671[4] can provide assistance but, ultimately, compliance with the Regulations is required. It is likely to be reasonably practicable to replace it with equipment made to a more recent standard when, but only when, it becomes unsafe or falls due for replacement for other than safety reasons, whichever occurs sooner.

British Standard BS 7671 Requirements for Electrical Installations (also known as the IET Wiring Regulations)

8   BS 7671 *Requirements for electrical installations* is also known as the IET Wiring Regulations. They are non-statutory regulations which 'relate principally to the design, selection, erection, inspection and testing of electrical installations, whether permanent or temporary, in and about buildings generally and to agricultural and horticultural premises, construction sites and caravans and their sites'.

9   BS 7671 is a code of practice which is widely recognised and accepted in the UK and compliance with it is likely to achieve compliance with relevant aspects of the Electricity at Work Regulations 1989.

10   There are, however, many types of system, equipment and hazard to which BS 7671 is not applicable; for example, certain installations at mines and quarries, equipment on vehicles, systems for public electricity supply and explosion protection. Furthermore, BS 7671 applies only to installations operating at up to 1000 V ac or 1500 V dc.

11   Installations to which BS 7671 is relevant may have been installed in accordance with an earlier edition, now superseded but then current. That, in itself, would not mean that the installation would fail to comply with the Regulations.

Electricity Safety, Quality and Continuity Regulations

12   The Electricity Safety, Quality and Continuity Regulations 2002 (ESQCR)[5] impose requirements in respect of the generation, distribution and supply of electricity, including supply networks and electrical equipment. The Department of Energy and Climate Change (DECC) leads on ESQCR matters, though HSE performs some functions on DECC's behalf, in particular with regards to public safety and incident notifications. Some ESQCR obligations – such as requirements for connection with earth – overlap with the Regulations, but others provide additional requirements.

The Regulations

Regulation 1 Citation and commencement

Regulation 1

These Regulations may be cited as the Electricity at Work Regulations 1989 and shall come into force on 1st April 1990.

Regulation 2 Interpretation

Regulation 2

(1) In these Regulations, unless the context otherwise requires –

*"circuit conductor" means any **conductor** in a **system** which is intended to carry electric current in normal conditions, or to be energised in normal conditions, and includes a combined neutral and earth conductor, but does not include a conductor provided solely to perform a protective function by connection to earth or other reference point;*

*"conductor" means a **conductor** of electrical energy;*

*"danger" means risk of **injury**;*

"electrical equipment" includes anything used, intended to be used or installed for use, to generate, provide, transmit, transform, rectify, convert, conduct, distribute, control, store, measure or use electrical energy;

"injury" means death or personal injury from electric shock, electric burn, electrical explosion or arcing, or from fire or explosion initiated by electrical energy, where any such death or injury is associated with the generation, provision, transmission, transformation, rectification, conversion, conduction, distribution, control, storage, measurement or use of electrical energy;

"system" means an electrical system in which all the electrical equipment is, or may be, electrically connected to a common source of electrical energy, and includes such source and such equipment.

Guidance 2

13 Words and phrases which are in **bold** type in the text of the regulation preceding the guidance are those which have been assigned a special meaning by being defined in regulation 2.

Systems

14 The term 'system' includes all parts of a system, eg conductors and electrical equipment in it, and is not a reference solely to the functional circuit as a whole. It follows that something required of a system is required both of the system as a whole and of the equipment and conductors in it.

Guidance 2

15 The definition refers to electrical systems. In the case of each system this will include all of the electrical equipment connected together and the various electrical energy sources in that system. In the case of transformers, even though there may be galvanic separation between the various windings of the transformers, where the energy is transmitted through these from one part of the electrical system to another, the transformer and all of its windings are part of the same system.

16 The definition of 'system' includes equipment which, although not energised, may be electrically connected to a common source of electrical energy. Equipment which is readily capable of being made live by a system is therefore considered to be part of that system. For example, a lighting circuit which has been disconnected from its source of electrical energy by means of removable links or fuses is still part of that system, as is such a circuit which has been switched off even though the switch might be a double pole switch.

17 Equipment which is in any way connected to a source of electrical energy, eg a test instrument containing a source and the equipment containing or connected to that source, becomes part of a system and the Regulations apply to that system. Electrical equipment which is not connected, and cannot be readily connected, to a source of electrical energy is not part of a system. Protective conductors, if they are connected to a source, are part of that system.

18 The reference in the definition to a common source of electrical energy does not exclude systems fed by several generators or transformers. The word 'common' is included in the definition so that completely independent electrical installations are regarded as separate systems. If, however, they are electrically connected in any way they are part of the same system for the purposes of the Regulations. This may mean that the system may be an extensive electrical network covering large geographical areas over which several or even many people have control of various parts. The Regulations place duties on these people only in respect of those provisions of the Regulations which relate to matters which are within their control (see regulation 3).

19 Self-contained portable systems, such as portable generating sets, are electrical systems for the purpose of the Regulations, as are transportable systems and systems on vehicles etc.

Electrical equipment
20 'Electrical equipment' as defined in the Regulations includes every type of electrical equipment from, for example, a high-voltage transmission overhead line to a battery-powered hand lamp. There are no voltage limits in the Regulations; the criteria are whether danger (as defined) may arise. It is appropriate for the Regulations to apply even at the very lowest end of the voltage or power spectrum because the Regulations are concerned with, for example, explosion risks, which may be caused by very low levels of energy igniting flammable gases even though there may be no risk of electric shock or burn.

21 Electrical equipment (as defined) includes conductors used to distribute electrical energy such as cables, wires and leads and those used in the transmission at high voltage of bulk electrical energy, as in the national grid.

Conductors

22 Regulation 2 defines a conductor as 'a conductor of electrical energy'. This means any material which is capable of conducting electricity (electricity is synonymous with electrical energy) and therefore includes both metals and all other

Guidance 2

conducting materials. The definition is not limited to conductors intended to carry current and so includes, for example, metal structures, salt water, ionised gases and conducting particles. The conductance of most materials varies with parameters such as temperature, eg glass is conducting when molten (and is then a conductor, as defined) whereas in its normal, solid, state it is a good insulator and finds many applications as such. For the purposes of the Regulations, while such materials conduct electricity, they are 'conductors'.

Figure 1 Types of conductor

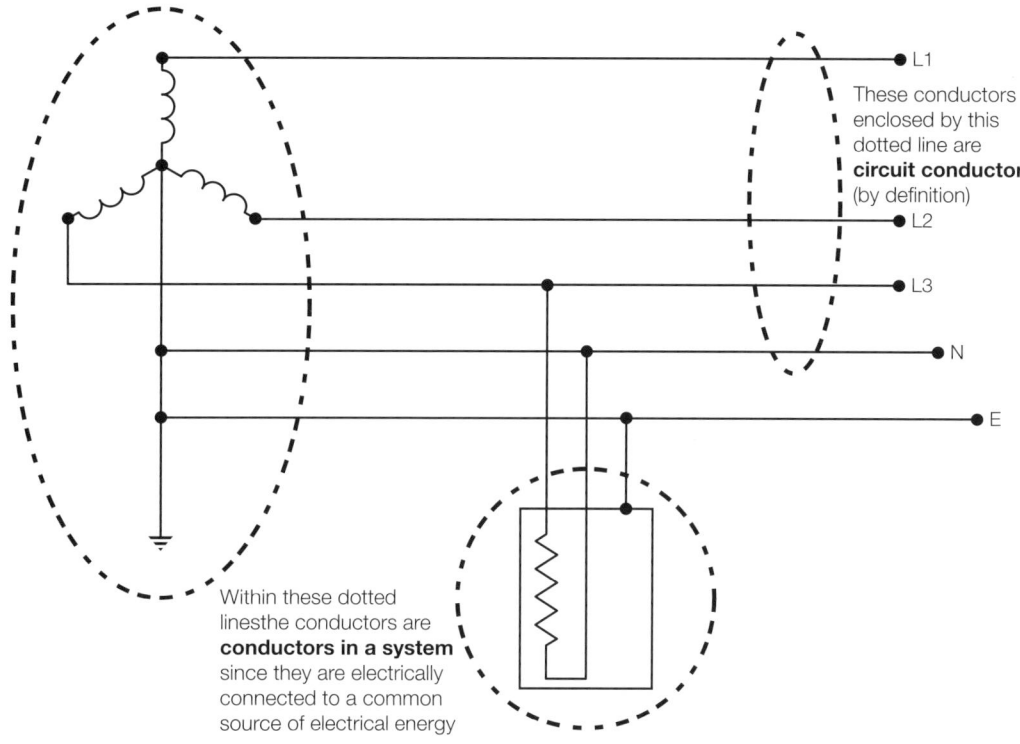

3 Phase system – separate neutral and earth

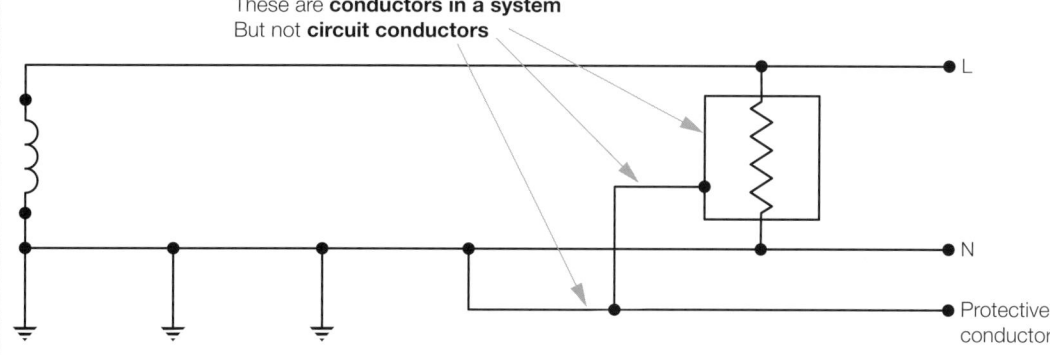

System including both combined and separate neutral and earth conductors (single phase only shown)

Circuit conductor

23 This definition is used in regulations 8 and 9 only. It distinguishes from all other conductors those conductors whose normal function is to carry load current or to be energised (see Figure 1).

Danger

24 The Regulations use the two defined terms, 'danger' and 'injury'. 'Danger' is defined as 'risk of *injury*'. 'Injury' is defined in terms of certain classes of potential harm to people.

25 Where the term 'prevent danger' is used it should therefore be read as 'prevent the risk of *injury*'.

26 The Regulations make requirements to 'prevent danger' or 'prevent injury' – or in the case of regulation 16 – 'to prevent danger or, where appropriate, injury'. The purpose of the distinction between 'injury' and 'danger' is to accommodate those circumstances when people must work on or so near live equipment that there is a risk of 'injury', ie where 'danger' is present and cannot be prevented. In these circumstances, under regulation 14, danger may be present but injury must be prevented.

27 The type of injuries with which the Regulations are concerned are detailed in the definition of 'injury' in the regulation (see paragraphs 29 and 30). The scope of the Regulations does not include consequential dangers such as crushing injuries caused by a machine going out of control following an electrical malfunction. Such other dangers are subject to other legal requirements under, for example, the HSW Act, the Factories Act 1961 and the Office, Shops and Railway Premises Act 1963.

28 If no danger arises from a particular system, item of electrical equipment or conductor and will not arise, then the Regulations, although applying to it, do not require any precautions to be taken. However, in order for there to be no danger, there would have to be no risk of electric shock, electric burn, fire, arcing or explosion.

Injury

29 The purpose of the Regulations is to prevent death or personal injury to any person from electrical causes in connection with work activities.

30 'Injury' means death or injury to any person from:

(a) electric shock;
(b) electric burn;
(c) fires of electrical origin;
(d) electric arcing;
(e) explosions initiated or caused by electricity.

Electric shock
31 The human body responds in several ways to electrical current flowing through it. The sensation of shock is only one such effect and this can be extremely painful. When a shock is received, the electric current may take multiple paths through the body and its intensity at any one point is difficult or impossible to predict. The passage of electric current may cause muscular contractions, respiratory failure, fibrillation of the heart, cardiac arrest or injury from internal burns. Any of these can be fatal.

Guidance 2

32 The nature and severity of injury depends upon the magnitude, duration and path of the current through the body and, in the case of alternating current (ac), on its frequency. It is not possible to identify precise thresholds for the existence of hazard because a judgement has to be made in each case taking all the circumstances into account, such as body weight, physical condition of the victim and so forth. Nevertheless, a guide to the sort of current magnitudes which mark the occurrence of various dangerous effects is given in the International Electrotechnical Commission's (IEC's) publication *Guide to effects of current on human beings and livestock. Special aspects relating to human beings*.[6] Quite low currents, of the order of only a few milliamps (mA), can cause fatal electric shock.

33 The likely effects of shock current are mainly influenced by:

(a) voltage;
(b) frequency;
(c) duration;
(d) any impedance in the current path.

34 The effects of electric shock are most acute at around the public electricity supply frequency of 50 hertz. Susceptibility to electric shock is increased if a person is in:

(a) good electrical contact with earth, such as in damp or wet conditions or in conducting locations such as inside a metal tank;
(b) hot environments where people may become damp due to perspiration or humidity, thus reducing the insulation protection offered by clothing.

35 The variability of conditions makes it impossible to specify a voltage which is guaranteed to be safe in all situations. In any situation, you should consider the risk of injury from electric shock against published guidance as to the voltages and other factors which have been found by extensive experience to be safe. This includes national and international standards and technical publications. However, these documents should be interpreted carefully and with a view to the limitations of their various scopes and assumptions. You should always consider the conventional public electricity supply voltage of 230 V ac as potentially fatally dangerous. Many fatal electric shock accidents have occurred from contact with conductors live at this voltage and possibly the most dangerous situation is where contact is made with conductors by each hand, current then flowing 'hand to hand' across the heart region.

36 The following documents give some guidance (see References):

(a) *Guide to effects of current on human beings and livestock* (IEC TS 60479);
(b) *Requirements for electrical installations* (BS 7671);
(c) *Safety of equipment electrically connected to a telecommunications network* (IEC 62151).[7]

Electric burn

37 Electric burns are different from burns due to fire (see paragraphs 39–40), arcing (see paragraphs 41–42) or explosion (see paragraphs 44–46). They are due to the heating effect caused by the passage of electric current through body tissues. They are most commonly associated with electric shock and often occur in and on the skin layers at the point of contact with the electrical conductors which gave rise to the electric shock. Electrical arcs jump across gaps and can cause burns.

Guidance 2

38   At high frequencies, eg radio frequencies (RF), which include microwaves, it may not even be necessary for contact to be made with live conductors for an electric burn to be received. In the case of RF, the heating is by absorption of the electromagnetic wave energy by a dielectric loss process in the body of the victim. RF burns can thus be extremely deep within the body. RF burning can occur without the sensation of shock, particularly if no contact is made with the RF conductors, and can therefore cause severe injury before the victim is aware of their occurrence. Electric burns are usually painful and very slow to heal. Permanent scarring is common.

Fires of an electrical origin

39   Fires may be started by electricity in a number of ways. The main causes are:

(a)   overheating of cables and electrical equipment due to overloading of conductors;
(b)   leakage currents due to poor or inadequate insulation;
(c)   overheating of flammable materials placed too close to electrical equipment which is otherwise operating normally;
(d)   the ignition of flammable materials by arcing or sparking of electrical equipment, including the scattering of hot particles from electrical equipment.

40   The injuries associated with fire are usually burns but may include other injuries such as smoke inhalation, each with the potential to kill.

Arcing

41   Arcing causes a particular type of burn injury which is distinct from other types. Arcing generates ultraviolet radiation which causes damage similar to severe sunburn. Molten metal particles from the arc itself can penetrate burn and lodge in the flesh. These effects are additional to any radiated heat damage caused by the arc.

42   On its own, ultraviolet radiation can cause damage; sensitive skin and eyes are especially vulnerable to arc flash. ('Arc eye' is commonly encountered with electric arc welding if the proper precautions are not adopted.)

43   Arcing faults can occur if the energy available at a piece of electrical equipment is sufficient to maintain a conductive path through the air or insulation between two conductors which are at different potentials. Under fault flashover conditions, currents many times the nominal rating or setting of a protective device may flow before those devices operate to clear the fault. Much energy is dissipated in the arc and, depending on the electrical protection, may continue long enough to inflict very serious arcing burns or to initiate a fire. These periods can be as short as 0.2 seconds. Arc flashovers caused during work on live circuit conductors are likely to be particularly hazardous because the worker is likely to be very near to or even enveloped by the arc. Such cases often lead to very serious, sometimes fatal, burn injuries.

Explosion

44   This category includes those injuries caused by explosions either of an electrical nature or those ignited by an electrical source.

45   Electrical explosions include the violent and catastrophic rupture of any electrical equipment. Switchgear, motors and power cables are liable to explode if they are subjected to excessive currents, which release violent electromagnetic forces and dissipate heat energy, or if they suffer prolonged internal arcing faults.

46   Explosions caused by ignition from an electrical source include ignition of flammable vapours, gases, liquids and dusts by electrostatic discharge, electric sparks, arcs or the high surface temperature of electrical equipment.

Guidance 2

Other words used in the Regulations

Charged/live (as used in regulations 8, 13 and 14)
47 The terms 'charged' and 'live' have different meanings; they are not defined in the Regulations so they take their ordinary meaning. 'Live' means equipment that is at a voltage by being connected to a source of electricity. 'Charged' means that the item has acquired a charge either because it is live or because it has become charged by other means, such as static or induction charging, or has retained or regained a charge due to capacitance effects even though it may be disconnected from the rest of the system.

Dead (as used in regulations 13 and 14)
48 The term 'dead' is not defined in the Regulations so it takes its ordinary meaning. Thus, in the context of the Regulations, for a conductor to be 'dead' means that it is neither 'live' nor 'charged'.

Other words used in the guidance

High voltage, low voltage
49 High voltage is a voltage in excess of 1000 V ac or 1500 V dc (direct current). Low voltage is a voltage up to and including 1000 V ac or 1500 V dc.

Live work
50 Live work is work on or near conductors that are accessible and 'live' or 'charged'. Note that testing of live exposed conductors using a test instrument is live work.

Regulation 3 Persons on whom duties are imposed by these Regulations

Regulation 3

(1) Except where otherwise expressly provided in these Regulations, it shall be the duty of every –

(a) employer and self-employed person to comply with the provisions of these Regulations in so far as they relate to matters which are within his control; and

(b) [(i) mine operator, in relation to a mine within the meaning of regulation 3 of the Mines Regulations 2014, and]*
(ii) operator, in relation to a quarry within the meaning of regulation 3 of the Quarries Regulations 1999,
to ensure that all requirements or prohibitions imposed by or under these Regulations are complied with in so far as they relate to the mine of which he is the [mine operator]* or quarry of which he is the operator and to matters which are within his control.

(2) It shall be the duty of every employee while at work –

(a) to co-operate with his employer so far as is necessary to enable any duty placed on that employer by the provisions of these Regulations to be complied with; and

(b) to comply with the provisions of these Regulations in so far as they relate to matters which are within his control.

[(3) In this regulation, "mine operator" has the meaning given by regulation 2(1) of the Mines Regulations 2014.]*

* Amended by the Mines Regulations 2014, SI 2014/3248.

Guidance 2

Guidance 3

Employer

51 For the purposes of the Regulations, an employer is any person or body who:

(a) employs one or more individuals under a contract of employment or apprenticeship; or
(b) provides training under the schemes to which the HSW Act applies through the Health and Safety (Training for Employment) Regulations 1990 (SI 1990/1380).

Self-employed

52 A self-employed person is an individual who works for gain or reward other than under a contract of employment, whether or not they employ others. They encounter the same risks as employed people.

Employee

53 Regulation 3(2)(a) reiterates the duty placed on employees by section 7(b) of the HSW Act.

54 Regulation 3(2)(b) places duties on employees equivalent to those placed on employers and self-employed people where these are matters within their control. This will include those trainees who will be considered as employees under the Regulations described in paragraph 51.

55 This arrangement recognises the level of responsibility which many employees in the electrical trades and professions are expected to take on as part of their job. The 'control' which they exercise over the electrical safety in any particular circumstances will determine to what extent they hold responsibilities under the Regulations to ensure that the Regulations are complied with.

56 A person may find themselves responsible for causing danger to arise elsewhere in an electrical system, at a point beyond their own installation. This situation may arise, for example, due to unauthorised or unscheduled back feeding from their installation onto the system, or to raising the fault power level on the system above rated and agreed maximum levels due to connecting extra generation capacity etc. Because such circumstances are within their control, the effect of regulation 3 is to bring responsibilities for compliance with the rest of the Regulations to that person, thus making them a dutyholder.

Absolute/reasonably practicable

57 Duties in some of the Regulations are subject to the qualifying term 'reasonably practicable'. Where qualifying terms are absent the requirement in the regulation is said to be absolute. The meaning of reasonably practicable has been well established in law. The interpretations in paragraphs 59–60 are given only as a guide to dutyholders.

Absolute
58 If the requirement in a regulation is 'absolute', for example if the requirement is not qualified by the words 'so far as is reasonably practicable', the requirement must be met regardless of cost or any other consideration. Regulations making such absolute requirements are subject to the defence provision of regulation 29.

Guidance 3

Reasonably practicable

59 Generally, you should do everything 'reasonably practicable' to protect people from harm. This means balancing the level of risk against the measures needed to control the real risk in terms of money, time or trouble. However, you do not need to take action if it would be grossly disproportionate to the level of risk.

60 In the context of the Regulations, where the risk is very often that of death, eg from electrocution, and where the nature of the precautions which can be taken are so often very simple and cheap, eg insulation, the level of duty to prevent that danger approaches that of an absolute duty.

Regulation 4 Systems, work activities and protective equipment

Regulation 4

(1) All **systems** shall at all times be of such construction as to prevent, so far as is reasonably practicable, **danger.**

(2) As may be necessary to prevent **danger**, **all systems** shall be maintained so as to prevent, so far as is reasonably practicable, such **danger**.

(3) Every work activity, including operation, use and maintenance of a **system** and work near a **system**, shall be carried out in such a manner as not to give rise, so far as is reasonably practicable, to **danger.**

(4) Any equipment provided under these Regulations for the purpose of protecting persons at work on or near **electrical equipment** shall be suitable for the use for which it is provided, be maintained in a condition suitable for that use, and be properly used.

Guidance 4

61 Regulation 4 covers, in a general way, those aspects of electrical systems and equipment, and work on or near these, which are fundamental to electrical safety.

Regulation 4(1)

62 The word 'construction' in the regulation has a wide application. It may be considered to cover the physical condition and arrangement of the components of a system at any time during its life. It will include aspects such as the design of the system and the equipment comprising that system.

63 In assessing the suitability of the construction of electrical systems, consideration should be given to all likely or reasonably foreseeable conditions of actual application or use of the electrical equipment in the system. This will include the testing, commissioning, operation and maintenance of the equipment throughout the life of the system.

64 In particular, you should consider:

(a) the manufacturer's assigned or other certified rating of the equipment;
(b) the likely load and fault conditions;
(c) the need for suitable electrical protective devices;
(d) the fault level at the point of supply and the ability of the equipment and the protective devices to handle likely fault conditions;
(e) any contribution to the fault level from the connected loads such as from motors;

> **Guidance 4**
>
> (f) the environmental conditions which will have a bearing on the mechanical strength and protection required of the equipment;
> (g) the user's requirements of the installation;
> (h) the risks that a system may create to adjacent work activities and the public;
> (i) the manner in which commissioning, testing and subsequent maintenance or other work may need to be carried out.
>
> 65 The safety of a system depends upon the proper selection of all the electrical equipment in the system and the proper consideration of the inter-relationship between the individual items of equipment. For example, electrical protection against overloads and earth faults etc may need to be provided in one part of a system to protect another, possibly remote, part of the system. Also, where electrical energy is transformed or converted from one voltage to another, you should take precautions to prevent danger arising from the lower-voltage conductors becoming charged above their normal voltage.
>
> ## Regulation 4(2)
>
> 66 Regulation 4(2) is concerned with the need for maintenance to be done to ensure safety of the system, rather than with the activity of doing the maintenance in a safe manner (which is required by regulation 4(3)).
>
> 67 The obligation to maintain arises only if danger would otherwise result. The maintenance should be sufficient to prevent danger so far as is reasonably practicable.
>
> 68 Inspection and, where necessary, testing of equipment is an essential part of any preventive maintenance programme. Practical experience of use may indicate an adjustment to the frequency at which preventive maintenance needs to be carried out. This is a matter for the judgement of the dutyholder, who should seek all the information they need to make this judgement including reference to the equipment manufacturer's guidance.
>
> 69 Records can aid demonstration of compliance and allow useful analysis of equipment condition, although keeping records is not a legal requirement. Maintenance records (including test results), preferably kept throughout the working life of an electrical system, will allow the condition of the equipment and the effectiveness of maintenance policies to be monitored. Without effective monitoring, dutyholders cannot be certain that the requirement for maintenance has been complied with.
>
> 70 British Standard Codes of Practice offering guidance on maintenance are referred to in Further reading. Advice on inspection and testing of some fixed installations is given in BS 7671.
>
> ## Regulation 4(3)
>
> 71 Regulation 4(3) requires that work activities of any sort, whether directly or indirectly associated with an electrical system, must be carried out in a way which, as far as is reasonably practicable, does not give rise to danger. Regulations 12 to 16 provide more specific requirements in connection with work of an electrical nature on or near electrical systems. See *Electricity at work: Safe working practices*[8] for more information

Guidance 4

Work activities associated with electrical systems

72 In the case of electrical work, it is preferable that the conductors are made dead before work starts (see regulations 12, 13 and 14). In such cases, it is essential that the equipment is isolated (note that 'isolation' is defined in regulation 12(2), which will include securing by locking off etc; see also paragraph 75) and the conductors proved dead at the point of work before the work starts. Where a test instrument or voltage indicator is used for this purpose, this device should itself be proved immediately before and immediately after testing the conductors.

73 Safe systems of work incorporating safety isolation procedures are important for work on equipment which is to be made dead before work starts. Particular consideration is needed where multiple sources of supply can exist, eg connection of alternative generation. These are also discussed under regulations 12 and 13. Some work, such as fault finding and testing or live jointing by the electricity supply industry, may require electrical equipment to remain energised during the work. In these cases, regulation 14 makes particular requirements and regulation 4(4) is also likely to be relevant in terms of the protective equipment which may need to be provided. *Electricity at work: Safe working practices* gives further guidance.

74 The operation, maintenance and testing of electrical systems and equipment must only be carried out by those people who are competent for that work (see also regulation 16).

Disused electrical equipment and systems

75 Before electrical equipment or systems are decommissioned they must be disconnected from all sources of supply and isolated. Similarly, electrical equipment or systems which are disused, or no longer required or abandoned for any reason, should be disconnected from all sources of supply and isolated. Isolation (as defined in regulation 12(2)) requires taking effective steps to ensure that it is dead and cannot become inadvertently re-energised or charged by induction or capacitance effects. (Regulations 12, 13 and 14 are also likely to be relevant.) Suitable labels or notices to bring people's attention to the state of the equipment are likely to be necessary in preventing inadvertent re-energisation.

Other work near electrical systems

76 Regulation 4(3) is wide in its application and includes work of a non-electrical nature where there is a risk of electrical injury. A common example is excavation near to live electric power cables and work near live overhead power lines, where the risks can be severe. Advice on these matters is given in HSE publications listed in References and further reading. You must also consider the requirements of regulation 14.

Regulation 4(4)

77 The defence (regulation 29) is available in any proceedings for an offence under this part of regulation 4.

78 The term 'protective equipment' can be of wide application but typically includes those special tools, protective clothing and insulating screening materials etc necessary to do work safely on live electrical equipment. The requirement for

Guidance 4

suitable precautions to prevent injury may arise under regulation 14. The regulation requires the protective equipment to be:

(a) suitable for use;
(b) maintained in that condition;
(c) properly used.

79 Regulation 4(4) is not qualified by 'so far as is reasonably practicable', nor does the regulation refer either to injury or the risk of injury, ie electrical danger. The impact of the regulation is that, where protective equipment is provided in order to comply with any of the other regulations, the equipment must conform to the requirements of regulation 4(4). Advice on safe working practices is given in HSE guidance. Specifications for certain types of protective equipment such as insulating gloves and floor mats are listed in relevant standards in References and further reading.

Regulation 5 Strength and capability of electrical equipment

Regulation 5

*No **electrical equipment** shall be put into use where its strength and capability may be exceeded in such a way as may give rise to **danger**.*

Guidance 5

80 The defence (regulation 29) is available in any proceedings for an offence under this regulation.

81 Before equipment is energised, the characteristics of the system to which the equipment is connected must be taken into account. This should include those existing under normal conditions, possible transient conditions and prospective fault conditions, so that the equipment is not subjected to stress which it is not capable of handling without giving rise to danger. The effects to be considered include voltage stress and the heating and electromagnetic effects of current.

Strength and capability

82 The term 'strength and capability' of electrical equipment refers to the ability of the equipment to withstand the thermal, electromagnetic, electrochemical or other effects of the electrical currents which might be expected to flow when the equipment is part of a system. These currents include, for example, load currents, transient overloads, fault currents, pulses of current and, for alternating current circuits, currents at various power factors and frequencies. Insulation must be effective to enable the equipment to withstand the applied voltage and any likely transient over-voltages.

83 A knowledge of the electrical specification and the tests, usually based on the requirements of national or international standards, will assist the user in identifying the withstand properties of the equipment so that it may be selected and installed to comply with this regulation. Such tests are normally carried out either by the manufacturer or by an accredited testing organisation.

Rating

84 The strength and capability of electrical equipment is not necessarily the same as its rating. Usually the rating is that which has been assigned by the manufacturer following a number of agreed tests.

Guidance 5

85 Electrical equipment should be used within the manufacturer's rating (continuous, intermittent or fault rating as appropriate) and in accordance with any instructions supplied with the equipment.

Fault conditions

86 So that equipment remains safe under prospective fault conditions, you must select equipment that takes account of the fault levels and the characteristics of the electrical protection which has been provided for the purpose of interrupting or reducing fault current (excess current protection is required by regulation 11). Most electrical equipment will be able to withstand short-circuit currents safely for limited periods only. The considerations also extend to conductors and equipment provided solely for protective purposes, eg earthing conductors must be adequately rated to survive beyond fault clearance times to ensure satisfactory protective gear operation and fault clearance.

Regulation 6 Adverse or hazardous environments

Regulation 6

***Electrical equipment** which may reasonably foreseeably be exposed to –*

(a) mechanical damage;
(b) the effects of the weather, natural hazards, temperature or pressure;
(c) the effects of wet, dirty, dusty or corrosive conditions; or
(d) any flammable or explosive substance, including dusts, vapours or gases,

*shall be of such construction or as necessary protected as to prevent, so far as is reasonably practicable, **danger** arising from such exposure.*

Guidance 6

87 The regulation draws attention to the kinds of adverse conditions where danger could arise if equipment is not constructed and protected to withstand such exposure. Electrical equipment must be suitable for the environment and conditions of use to which it may reasonably foreseeably be exposed so that danger which may arise from such exposure will be prevented so far as is reasonably practicable. The following paragraphs detail some of the conditions to which electrical equipment may be subjected. Guidance is given in these paragraphs and additional guidance may be found in the documents listed in References and further reading. Particular attention should be paid to the IP rating (Index of Protection) of equipment (see paragraph 108). Guidance is also given under regulation 8 on the use of reduced voltage systems on construction sites and elsewhere where particularly arduous or conducting locations may exist (see paragraphs 142–144).

Effects

88 The conditions at which the regulation is directed are those occurring naturally as well as those resulting from human activities, including the following:

(a) mechanical damage including impact, stress, strain, abrasion, wear, vibration and hydraulic and pneumatic pressure;
(b) effects of the weather, which include both short-term (eg wind, ice and snow, lightning) and long-term (eg temperature cycling) effects;
(c) natural hazards, which are those resulting from other than man's activities and include animals, trees and plants, tides and solar radiation etc;
(d) temperature and pressure;

Guidance 6

(e) liquids which include water and other liquids and their effects, including humidity, condensation, flooding, splashing, or immersion in these, cleaning with liquids, hosing down and solvent and solvent vapour action (electrically conducting and non-conducting liquids may present different aspects of electrical danger);

(f) dirty conditions which include all contamination as a result of liquids or solids (electrically conducting and non-conducting dusts may present different aspects of electrical danger);

(g) corrosive conditions which include all chemical actions and reactions and electrochemical effects;

(h) flammable substances, including flammable dusts and flammable vapours;

(i) explosive substances which include both any mixture of solids, liquids or gases which is capable of exploding and substances intended to be explosive (ie explosives).

89 When determining the suitability of equipment for particular environments or conditions of use, you only need to consider exposure or effects which are reasonably foreseeable.

Mechanical damage

90 The mechanical damage to which electrical equipment may be subjected varies considerably from one environment to another. For example, equipment designed for use in an office is unlikely to be suitable, without further protection or careful siting, in a workshop or farm environment.

91 The effects covered by regulation 6(b), (c) and (d) may also impose mechanical stresses on electrical equipment. For example, ice and wind loading, or loss of mechanical strength due to expansion and contraction resulting from temperature changes, can give rise to mechanical damage.

92 This regulation requires the mechanical protection, if necessary, of the insulation which is required under regulation 7(a). For example, to protect against impact damage, steel wire armouring of a cable may be necessary. Further suitable protection in addition to basic insulation may be required to form the physical protection necessary to ensure the continuing integrity of basic insulation, eg conduits or a trunking for single-insulated conductors or the armouring or tough external sheathing of composite or multi-core electric cable.

Weather, natural hazards and extreme conditions

93 Precautions taken to protect a site, structure or building from natural hazards and extreme weather conditions may give some protection to the associated electrical installation, but additional protection or precautions may also be necessary.

94 Extremes of temperature, pressure or humidity may result either from climatic conditions or from adjacent plant or from the use of the electrical equipment itself. Standards frequently quote the range of service conditions for electrical equipment, including temperature limits, and users should consider these when selecting equipment.

95 Guidance on assessing the need for lightning protection of structures and buildings etc, the design and provision of systems and their inspection, testing and maintenance is given in publications listed in References and further reading.

Guidance 6

Corrosive effects

96 If substances are present in the environment that accelerate corrosion of metallic enclosures or fittings, special materials or surface treatments may be necessary. In these cases, the electrical equipment must be protected, eg motors, and be of a type which is totally enclosed by an appropriate corrosion-resistant housing, ie not ventilated to the atmosphere.

97 Insulating materials and other materials used in electrical equipment may be affected by chemical agents or solvents. Cubicles housing electrical control equipment in hostile environments may need to be kept purged or pressurised with clean air or, in special cases, inert gas. See Further reading for details of relevant standards.

Dirt and dusts

98 Most industrial enclosures for electrical equipment do not resist the entry of fine dusts. Equipment must be constructed to resist the entry of dust and dirt where this may give rise to electrical and mechanical failures. Appropriate regular inspection and cleaning are recommended where dirt and dusts are likely to accumulate, eg portable motor-driven equipment incorporating ventilation slots which can cause the accumulation of potentially hazardous layers of dirt and dust.

Combustible dusts

99 In cloud form, some dusts create an explosion hazard, while layers of combustible dust on electrical equipment can give rise to fire hazards. The selection, construction or installation of equipment exposed to combustible dust must guard against the possibility of ignition. The maximum temperature attainable on the surface of any electrical equipment where these dusts may be deposited should always be below the temperature at which any ignition, charring or smoking of dust takes place. However, appropriate dust control measures and general cleanliness which minimise the problem at source are recommended. See References and further reading for details of relevant standards.

Potentially explosive atmospheres

100 If electrical equipment is used where a flammable or explosive atmosphere is likely to occur, the equipment must be constructed so that it is not liable to ignite that atmosphere. Further information is available in the Dangerous Substances and Explosive Atmospheres Regulations.[9]

101 The selection and installation of equipment for use in potentially explosive atmospheres should be guided by the recommendations contained in the HSE guidance and British Standards on the subject. Existing installations complying with the recommendations of earlier standards should be acceptable for continuing service, subject to proper maintenance.

102 Such electrical equipment must be chosen from that which has been certified as conforming to an appropriate standard.

103 Uncertified electrical equipment must not be used unless it will provide at least an equivalent level of safety to that provided by appropriately certified equipment.

Guidance 6

104 Some manufacturing processes, eg electrostatic paint spraying, make use of the characteristics of static electricity and the design of electrical equipment should be such that the ignition of solvents, vapours or particulate substances is prevented. See References and further reading for details of relevant standards.

105 The maintenance and repair of explosion-protected equipment is a specialised field of work and must only be carried out by those who have the necessary training and experience.

Other flammable substances

106 Electrical equipment that generates heat or produces sparks must not be placed where either the heat emitted or sparking could lead to the uncontrolled ignition of any substance.

107 The construction of the equipment should either exclude the substances from any part of the equipment which may be a source of ignition (eg by suitable enclosure) or should ensure that the equipment operates at sufficiently low temperature and energy levels as not to be a source of ignition under likely conditions of use and fault.

Classification system of ingress protection (IP rating)

108 There is an internationally recognised system of classifying the degree of protection provided by enclosures against the ingress of solid objects and moisture, and the protection afforded against contact with any live parts within the enclosure for all types of electrical equipment. The system is commonly known as the IP rating system and is detailed in a number of standards which are listed in Further reading.

Regulation 7 Insulation, protection and placing of conductors

Regulation 7

*All **conductors** in a **system** which may give rise to danger shall either –*

*(a) be suitably covered with insulating material and as necessary protected so as to prevent, so far as is reasonably practicable, **danger**; or have such precautions taken in respect of them (including, where appropriate, their being suitably placed) as will prevent, so far as is reasonably practicable, **danger**.*

Guidance 7

109 The danger to be protected against generally arises from differences in electrical potential (voltage) between circuit conductors or between such conductors and other conductors in a system – usually conductors at earth potential. The conventional approach is either to insulate the conductors or place them so people are unable to receive an electric shock or burn from them.

110 Some form of basic insulation, or physical separation, of conductors in a system is necessary for the system to function. That functional minimum, however, may not be sufficient to comply with the requirements of regulation 7. Factors which must be taken into account are:

Guidance 7

(a) the nature and severity of the probable danger;
(b) the functions to be performed by the equipment;
(c) the location of the equipment, its environment and the conditions to which it will be subjected;
(d) any work which is likely to be done on, with or near the equipment.

Insulation

111 Regulation 7(a) states that conductors must be insulated. Suitable insulation of the conductors in an electrical system is, in the majority of cases, the primary and necessary safeguard to prevent danger from electric shock, either between live conductors or between a live conductor and earth. It will also prevent danger from fire and explosion arising from contact of conductors either with each other or with earth. Energy from quite low levels of voltage (and levels insufficient to create a shock risk) can ignite a flammable atmosphere. The quality and effectiveness of insulation therefore needs to be commensurate with the voltages applied to the conductors and the conditions of use.

112 BS 7671 gives some advice on these matters for fixed electrical installations up to 1000 V ac or 1500 V dc.

113 The regulation requires that the insulation be protected as necessary, so that danger may be prevented so far as is reasonably practicable. Often, the protection required is to prevent mechanical damage to the insulation but may include any of the effects detailed under regulation 6. Examples of such protection would be the use of steel trunking and conduits or the use of steel armoured cables.

Other precautions including placing

114 Regulation 7(b) permits the alternative of having such precautions taken in respect of the conductors. These precautions may include the suitable placing of conductors. They may comprise strictly controlled working practices reinforced by measures such as written instructions, training and warning notices etc. The precautions must prevent danger so far as is reasonably practicable. Examples where bare conductors are used in conjunction with suitable precautions are to be found in many applications including overhead electric power lines, down-shop conductors for overhead travelling cranes in factories etc, railway electrification using either separate conductor and running rails or overhead pick-up wires, and certain large electrolytic and electrothermal plants.

115 The placing of overhead electric power lines is specified in ESQCR (see Introduction for more information).

116 Electric railway and tramway operators, in conjunction with the Office of Rail and Road, have developed standards and safety specifications for the construction of those parts of their systems which use bare conductors at overhead and at track level, together with safe systems of work.

117 Safety is ensured in electrochemical plants which use high current by such means as the separation of conductors which are at different potentials, the use of insulating working platforms and unearthed or isolated electrical supplies (see paragraphs 121–123).

118 Suitable placing of the conductors may, on its own, go a considerable way towards preventing danger, for example where the conductors are within a secure enclosure or where they are placed overhead at such a height that contact with

| Guidance | 7 |

these conductors is not reasonably foreseeable. Guidance on the security and protection of enclosures and the measure of their accessibility as determined by standard (finger) tests is given in standards listed in Further reading.

119 However, if the placing of the conductors cannot alone be relied upon to prevent danger, then additional precautions must be taken and rigorously applied. For example, in the case of live railway conductor rails the precautions may include warning notices, barriers and special training for railway staff. Electrolytic and electrothermal processes are further examples and are covered in paragraphs 121–123.

120 Dutyholders should carefully consider the inherent risks that may still exist if bare conductors are placed where they cannot normally be touched, eg maintenance activities around the conductors of an electric overhead crane system. Firstly, the protection of the equipment is required under regulation 6 for a range of reasonably foreseeable effects and, secondly, there may be occasions when people will require access to the area or enclosure where such conductors are located, eg substations and test areas. Where work is to be done with the conductors live, regulation 14 is relevant and the guidance under that regulation also applies.

Electrolytic and electrothermal processes

121 It is often necessary for industrial electrolytic and electrothermal processes, including large secondary battery installations, to adopt a range of precautions. As the work activity is likely to be near the live and uninsulated conductors, the precautions adopted will go towards satisfying both part (b) of regulation 7 and regulation 14.

122 Precautions may include:

(a) segregating the process area and limiting access to those people who are trained and experienced in the process and to people who are supervised so that injuries are prevented;
(b) ensuring a separation of conductors appropriate to the difference in potentials;
(c) use of insulating work platforms;
(d) use of electrical supplies which are isolated from earth together with protective devices to ensure this isolation;
(e) exclusion of unnecessary conducting materials and implements from the process area;
(f) use of protective clothing, eg in electric arc welding processes; such clothing offers protection against both the hot welding process and against the electric shock risk.

123 Details of advice on the safe use of electric induction furnaces and electric arc welding are given in Further reading.

Regulation 8 Earthing or other suitable precautions

Regulation 8

*Precautions shall be taken, either by earthing or by other suitable means, to prevent **danger** arising when any **conductor** (other than a **circuit conductor**) which may reasonably foreseeably become charged as a result of either the use of a **system**, or a fault in a **system**, becomes so charged; and, for the purposes of ensuring compliance with this regulation, a **conductor** shall be regarded as earthed when it is connected to the general mass of earth by **conductors** of sufficient strength and current-carrying capability to discharge electrical energy to earth.*

Guidance 8

124 The defence (regulation 29) is available in any proceedings for an offence under this regulation.

125 The regulation applies to any conductor (other than circuit conductors), including the conductive parts of equipment, such as outer metallic casings, which can be touched and, though not live, may become live under fault conditions.

126 Conductors which, although not part of a system, are within electrostatic or electromagnetic fields created by a system may be subject to this regulation. Appropriate precautions are necessary if the induced voltages or currents are large enough to give rise to danger.

Dangers

127 Dangers which may arise as a result of failure to take the necessary precautions include:

(a) risk of shock from conductors which are or may be exposed so that they may be touched and which become charged at dangerous voltage relative to earth or to other exposed conductors;
(b) risk of burns, fire, arcing or explosion due to currents of excessive magnitude and/or duration in such conductors.

128 The requirements of the regulation may be met in several different ways, depending on the circumstances, including:

(a) ensuring that such conductors do not become charged. This has the effect of excluding the conductors from the scope of this regulation;
(b) ensuring that if such conductors do become charged the values of voltage and current and the duration are such that danger will not arise;
(c) ensuring that if such conductors do become charged the environment is such that danger will not arise.

129 Techniques employed for achieving the above include:

(a) double insulation;
(b) earthing;
(c) connection to a common voltage reference point on the system;
(d) equipotential bonding;
(e) use of safe voltages;
(f) earth-free, non-conducting environments;
(g) current/energy limitation;
(h) separated or isolated systems.

130 You may employ the above techniques singly or in combination.

> **Guidance 8**

Double insulation

131 The principle of 'double insulation' is that the live conductors of the electrical equipment are covered by two discrete layers or components of insulation, each of which would adequately insulate the conductor but which together ensure an improbability of danger arising from insulation failure. This arrangement avoids the need for any external metalwork of the equipment to be connected to a protective conductor or to earth. Double insulation has been found to be particularly suitable for certain types of portable equipment, eg electric motor-driven tools etc. See Further reading for details of relevant standards. However, the integrity of this safety protection depends upon the layers of insulation remaining in sound condition and this in turn requires the equipment to be properly constructed, used and maintained.

Earthing

132 It is the practice in the UK for the public electricity supply system at the usual distribution voltages of 230 V single-phase, 400 V three-phase, to be referenced to earth by a deliberate electrical connection made at the distribution substations or power transformers. It is the existence of this system earthing which enables earth faults on electrical equipment to be detected and the electrical supply to faulty equipment to be cut off automatically.

133 Many 230/400 V power installations are designed so that the automatic interruption of the supply in the event of an earth fault is achieved by the operation of fuses or automatic circuit breakers (MCBs etc). In most cases, these devices will have been selected to provide the additional protective function of interrupting excess current required under regulation 11. In these circumstances the earth fault current must be large enough to rupture the fuse quickly. The magnitude of the fault current under full earth fault conditions is governed mainly by the combined impedance of the fault loop, which will include the impedance of the fault itself, that of the earthing or protective conductors, the circuit conductors and that of the source.

134 Tests must therefore be carried out on new installations and at appropriate intervals thereafter to ascertain that the earth fault (loop) impedances are low enough to ensure the electrical protective devices such as fuses, circuit breakers etc will operate in the event of a breakdown of insulation leading to an 'earth fault'. This includes the temporary electrical system used at a construction site in the process of carrying out construction activities.

135 Acceptable values of earth loop impedance and interruption times etc, for final installations up to 1000 V, can be found in BS 7671. It is rarely sufficient to rely on an earth rod or rods to provide sufficient conductance for return fault currents. Separate protective earth cables or conductors connected to the neutral point of the supply are usually necessary unless other measures, such as the use of sensitive residual current protection equipment, are used to detect earth fault currents.

136 For the duration of the fault, the electrical bonding of exposed conductive parts and their connection to earth serves to limit the shock risk from the transient voltages appearing between metallic enclosures of equipment in the system or between a metallic enclosure and earth. Equipment earthing therefore includes the bonding of metallic enclosures, cable armouring, conduits and trunking etc, so that these conductors are electrically continuous and securely connected to the general mass of earth at one or more points.

137 Earthing and bonding conductors must be suitable for the maximum current which they may carry under fault conditions and be capable of surviving the worst-

case fault (see paragraph 82). Their construction and strength must be adequate to withstand likely wear and tear. Where it might otherwise be difficult to ensure the continued effectiveness of earthing and bonding arrangements, it may be necessary to provide supplementary protection such as protective earth conductor monitoring.

138 Accidents have been caused by the metalwork of portable or transportable equipment becoming live as a result of the combined effects of a fault and high-impedance, protective conductor connections. The danger may be reduced by the use of a residual current device (RCD) designed to operate rapidly at small leakage currents (typically not exceeding 30 mA), although these devices do not eliminate the risk of electric shock. RCDs should not be considered as the sole means of protection but as an additional protective measure. They should be operated regularly using the test trip button. This test trip procedure is important in maintaining the effectiveness of most types of RCD.

139 Electric arc welding brings special problems associated with earthing practices. Stray currents from electrical arc welding can damage the protective earthing conductors of electrical installations. Information on earthing practice is available in a number of publications, some of which are listed in References and further reading.

Connection to a common voltage reference point on the system
140 In UK public electricity supply systems where transformer neutral points are connected to earth, the voltage reference point is the general mass of earth. Other reference points, to which systems may be referenced and to which bonding conductors are connected, may be chosen to suit particular circumstances.

Equipotential bonding
141 Equipotential bonding is the electrical interconnection of all exposed and extraneous conductors, which may become electrically charged, in such a way that dangerous voltages between any of the conductors that may be simultaneously touched are limited.

Use of reduced voltages
142 Reduced voltage systems are particularly appropriate for portable and transportable equipment, and in highly conducting locations such as boilers and tunnels where the risk of mechanical damage to equipment and trailing cables is high, and/or the body may be damp and have large areas of contact with the conducting location and on construction sites.

143 One example is a building or construction site supply system operating at 55-0-55 V ac single-phase, or at 110 V three-phase with a phase-earth voltage of 64 V ac. Another example is an extra-low-voltage system operating at or below 50 V ac or 120 V dc. Supply systems like these are referenced to earth and are therefore a special case of systems operating at reduced voltage for which bonding and earthing of all metallic enclosures are still recommended.

144 Further advice on reduced voltage systems is in Further reading.

Earth-free, non-conducting environments
145 If a system is supplied from a source which is earth-referenced, the path for fault current and the existence of dangerous potentials to earth can be eliminated in a defined area by ensuring that the area is 'earth-free'. This does not necessarily mean that metallic components or fittings need to be prohibited but rather that no part of the defined area is earthed. It is easier to ensure the integrity of an 'earth-free' area by constructing it from non-metallic components, in which case it is more appropriately known as a non-conducting location or area. 'Earth-free' and 'non-

Guidance 8

conducting' areas are specialised applications and are used mainly in certain testing of electrical equipment. Advice is available in the publications on electrical testing listed in References and further reading.

Current limitation

146 If fault currents which could cause electric shock are inherently limited by appropriate passive devices, eg high-integrity resistors, then protection by earthing or other means may not be required. In a conventional, dry working environment, for example, if the current is limited preferably to 1 mA but certainly to no more than 3 mA, this will not usually present a risk of injury from electric shock to people in good health who may be subjected to it only occasionally and for a short time. However, even this low level of current may give perceptible shock which, although by itself is unlikely to be physiologically dangerous, may give rise to a consequential injury such as from a fall induced by the shock. See paragraphs 29–30 on 'injury' under regulation 2, and especially IEC publication *Guide to effects of current on human beings and livestock*.

Separated or isolated systems

147 If safety depends on the supply system not being referenced to its immediate environment, whether true earth or surrounding metalwork, no potential should normally exist between live conductors and earth or exposed metallic parts. However, all systems are to some extent referenced to their environment by capacitive or inductive coupling or by leakage. That is why you cannot necessarily rely on the circuit conductors of separated or isolated* systems being at zero potential relative to their environment. Unless the isolated system is a very small and localised one, the leakage current may be large enough to provide a path for a fatal electric shock. Any difference in potential is likely to be greatest on extensive systems but, in all cases when the voltages or currents could be dangerous, precautions are needed. Examples of isolated systems are those supplied from the secondary winding of an isolating transformer or the winding of an alternator where there is no connection between them and any other source of electrical energy.

148 The isolation of a power system from earth may reduce the risks associated with a single fault. However, if this first fault has the effect of referencing the system to earth or other exposed conductor, subsequent faults may lead to very destructive and hazardous short circuits so extra precautions will be necessary to prevent this danger. These may include the bonding of metallic enclosures, earth fault detection, insulation monitoring or the use of an earth-free non-conducting environment. Regular inspection and testing to ensure that system isolation integrity is maintained will also be necessary.

* 'Isolated' in this context means separate from all other systems and does not imply 'isolation' as defined specifically for the purpose of regulation 12.

Regulation 9 Integrity of referenced conductors

Regulation 9

*If a **circuit conductor** is connected to earth or to any other reference point, nothing which might reasonably be expected to give rise to **danger** by breaking the electrical continuity or introducing high impedance shall be placed in that conductor unless suitable precautions are taken to prevent that **danger**.*

Guidance 9

149 The defence (regulation 29) is available in any proceedings for an offence under this regulation.

150 In many circumstances the reference point is earthed because the majority of power distribution installations are so referenced by a deliberate connection to earth at the generators or distribution transformers.

151 The object of the regulation is to prevent referenced circuit conductors which should be at or about the same potential as the reference point from reaching significantly different potentials, thereby giving rise to possible danger.

152 The most common situation in which this regulation is relevant is in systems having a neutral point which is earthed. Such systems can be subdivided:

(a) systems or parts of systems in which the neutral and protective conductor are combined (eg TN-C and the combined parts of TN-C-S systems);*
(b) systems or parts of systems in which the neutral and protective conductors are separate (eg TN-S and the separate parts of TN-C-S systems).

Devices placed in the conductor

153 The regulation does not prohibit all electrical devices from being placed in referenced conductors. For example, a proper joint or a bolted link or a bar primary current transformer can be arranged to ensure the integrity of the conductor.

154 The regulation would also permit the inclusion of other devices such as a removable link, or even a manually-operated knife switch, provided that suitable precautions are adopted to ensure that these devices are not removed or operated in such a way as to give rise to danger. However, a number of other devices, such as fuses, thyristors, transistors etc, generally have the potential to give rise to danger by becoming open circuit or introducing high impedance into the conductor. The regulation prohibits such applications.

Combined neutral and protective conductors

155 Open circuit of, or high impedance in, combined neutral and protective conductors will almost certainly result in the exposed and extraneous conductors connected to the protective conductors, eg metal enclosures of switchgear, being at a significant potential (up to phase-neutral volts) relative to earth. This could lead to a risk of electric shock or burn, so the integrity of the combined neutral and earth conductor is very important.

156 However, where the protective conductor is combined with the neutral conductor over some part of their length, you should take precautions to prevent people coming into simultaneous contact with the protective conductors and earth (or conductors at earth potential). Equipotential bonding of all metalwork within a building and the connection of this to the protective conductor or neutral is a

* This terminology is explained in BS 7671.

commonly used approach. Generally, however, combined neutral earth (CNE) systems should be confined to the public electricity supply network up to the point of supply to consumers.

Separate neutral and protective conductors

157 In deciding whether danger may result where there are separate neutral and protective conductors, consider not only the normal operation of the system but also the situations that may arise when work is being carried out on or near the system. If voltage rises on the neutral conductor could result in danger during the work then observe the above restrictions on devices in the neutral. For example, a fuse should not be placed in a neutral of a fixed power distribution installation (typically 230 V) as this places people working on the installation at risk of electric shock and burn if that fuse operates or otherwise becomes open circuit. Double pole fusing (fuses in both the phase and neutral) is acceptable, if these are fitted within self-contained electrical equipment which itself is not part of the fixed electrical installation, and is connected to the fixed installation by a plug and socket. This arrangement allows the equipment to be readily isolated from the system before work is done on that equipment.

158 In general, if a neutral conductor is to be switched, a multipole switch or circuit breaker should be used which also switches all of the related phase conductors, the neutral breaking last and making first. Such switching should not interrupt the protective conductor.

Regulation 10 Connections

*Where necessary to prevent **danger**, every joint and connection in a **system** shall be mechanically and electrically suitable for use.*

159 The defence (regulation 29) is available in any proceedings for an offence under this regulation.

Suitability of connections

160 The regulation requires that all connections in circuit and protective conductors, including connections to terminals, plugs and sockets, and any other means of joining or connecting conductors, should be suitable for the purposes for which they are used. This requirement applies equally to temporary and permanent connections. The insulation and conductance of the connections must be suitable, having regard to the conditions of use including likely fault conditions.

161 The mechanical protection and strength must be such as to ensure the integrity of the insulation and conductance under all conditions of use including likely fault conditions, subject to the need for any maintenance which may be required by regulation 4(2).

162 Joints and connections in protective conductors must be made at least as carefully as those in circuit conductors and they should be of sufficient strength and conductance to allow for the passage of fault currents. Such connections may need to be treated so they prevent corrosion. It is recommended that combinations of metals liable to produce damaging electrolytic action be avoided.

Guidance 10

Plugs and sockets

163 Plug and socket connections and their use must be arranged so that accidental contact with conductors live at dangerous voltages is prevented. This should be achieved by selection of appropriate equipment but may involve some degree of operator skill and/or training, depending on the circumstances.

164 In most applications, where a plug and socket type connector conveys a protective conductor as well as the circuit conductors, the protective conductor should be the first to be made and the last to be separated. The use of equipment made to appropriate standards should ensure this principle is adhered to.

165 Where plug and socket connections are not rated for making or breaking the maximum load current, effective arrangements should be made, eg by mechanical interlocking with the switch that controls the power, to ensure that the connections are made or broken only under no-load conditions.

Portable equipment

166 Special attention should be given to joints and connections in cables and equipment which will be handled, eg flexible cables for portable equipment. Plugs and sockets for portable equipment must be constructed in accordance with appropriate standards and arranged so that, where necessary, earthing of any metal casing of the equipment is automatically effected by the insertion of the plug. HSE guidance (see *Maintaining portable electric equipment in low-risk environments*[10]) and British Standards give further guidance on portable equipment.

Regulation 11 Means for protecting from excess of current

Regulation 11

*Efficient means, suitably located, shall be provided for protecting from excess of current every part of a **system** as may be necessary to prevent **danger**.*

Guidance 11

167 The defence (regulation 29) is available in any proceedings for an offence under this regulation (see paragraphs 177–179).

168 It is recognised that faults and overloads may occur on electrical systems. The regulation requires that systems and parts of systems be protected against the effects of short circuits and overloads if these would result in currents which would otherwise result in danger.

169 The means of protection is likely to be in the form of fuses or circuit breakers controlled by relays etc, or it may be provided by some other means capable of interrupting the current or reducing it to a safe value.

The need to anticipate abnormal conditions

170 The regulation requires the means of preventing danger to be provided in anticipation of excess current; a fault or overload need not have occurred. Fault currents arise as a result of short circuits between conductors caused either by inherent failure of the electrical equipment or some outside influence, eg mechanical damage to a cable. Overload currents can arise as a result of the inadequacy of a system to supply the load and may be caused by an increased demand created by

Guidance 11

outside influence on the electrical equipment, eg mechanical overloading of an electric motor.

The selection of excess current protection

171 In principle, every main circuit should be protected at its origin, ie at the source end of the circuit. Where the rating of the conductors forming a branch circuit is less than that of the conductors from which it is drawing power, it is conventional for protection to be placed at this point. In practice, however, there are exceptions to this principle and, depending on the nature of the system, a technical judgement must be made as to where the protection should be placed. Guidance on some aspects of this subject is given in BS 7671.

172 When selecting the means of protection, you must consider a number of factors – the more important of these include:

(a) the nature of the circuits and type of equipment to be protected;
(b) the short-circuit energy available in the supply (the fault level);
(c) the nature of the environment;
(d) whether the system is earthed or not.

The nature of the circuits and type of equipment to be protected

173 The circuits to be dealt with may vary from high-power, high-voltage circuits, eg for the inter-connection of substations or for the supply to large motors, down to the smallest final circuit supplying a few low-power lamps at, say, 6 V. Over this range lies a great diversity of equipment, each item of which will possess characteristics which must be carefully considered in the selection of appropriate devices to protect against excess current.

Fault level

174 The maximum short-circuit current in the protected circuit must be considered. (The ability of circuit breakers and fuses to operate successfully and without dangerous effects, serious arcing or, in the case of oil-filled equipment, the liberation of oil, is implicit in the requirements of regulations 4 and 5.) The design of the protective arrangement must also provide for sufficient current to be available to operate the protective devices correctly in respect of all likely faults.

The nature of the environment

175 The nature of the environment may have a bearing on the choice of protective devices and their settings, eg where the possibility of a fire being started may be considerable. However, in all cases, the protection against excess current must be effective so that short circuits and earth faults are cleared promptly to minimise destructive arcing and heating. Protective devices, whether they are circuit breakers or fuses, should therefore be set or selected for the minimum tripping currents and times consistent with ensuring the reliable operation of the device and with the need to discriminate between successive stages of protection.

Earthed system

176 Where a system is earthed, the nature and efficiency of the earthing system is important in relation to the design and reliability of the protective devices. In earthed systems, operation in the event of an earth fault of the protective device is dependent on sufficient current passing to operate the excess current or earth leakage tripping device or to blow the fuse. In many systems, the device provided to comply with the requirements of this regulation in respect of excess of current (very often a fuse) may also provide protection against earth faults – and thus be in compliance with the requirements of regulation 8.

| Guidance | 11 |

Defence in criminal proceedings

177 The defence (regulation 29) is available in any proceedings for an offence under this regulation.

178 In some circumstances it will be technically impossible to achieve total compliance with the absolute requirement to prevent danger. If an excess of current is drawn due to a fault or overload, eg due to an arcing fault, then whatever form of electrical protection is provided, there will be some danger at the point of the fault during the finite time taken for the detection and interruption of the fault current. Nevertheless, electrical protection – whether by means of a simple fuse or another method – must be properly chosen and installed in accordance with good electrical engineering practice. The protection must be efficient and effective.

179 In some circumstances it is undesirable to interrupt the current in a circuit because this may itself lead to a hazard. Examples include the excitation field current of direct current motors, trip coil circuits, lifting electromagnets and the secondary circuits of current transformers. In such cases, however, the circuit should be rated or arranged so as not to give rise to danger from excess of current.

Regulation 12 Means for cutting off the supply and for isolation

| Regulation | 12 |

(1) Subject to paragraph (3), where necessary to prevent **danger**, suitable means (including, where appropriate, methods of identifying circuits) shall be available for –

(a) cutting off the supply of electrical energy to any **electrical equipment**; and
(b) the isolation of any **electrical equipment.**

(2) In paragraph (1), "isolation" means the disconnection and separation of the **electrical equipment** from every source of electrical energy in such a way that this disconnection and separation is secure.

(3) Paragraph (1) shall not apply to **electrical equipment** which is itself a source of electrical energy but, in such a case as is necessary, precautions shall be taken to prevent, so far as is reasonably practicable, **danger.**

| Guidance | 12 |

180 The defence (regulation 29) is available in any proceedings for an offence under this regulation.

Regulation 12(1)(a)

181 The objective of this part of the regulation is to ensure that, where necessary to prevent danger, suitable means are available to switch off the electricity supply to any piece of equipment. Switching can be, for example, by direct manual operation or by indirect operation via 'stop' buttons in the control circuits of contactors or circuit breakers. There may be a need to switch off electrical equipment for reasons other than preventing electrical danger but these considerations are outside the scope of the Regulations.

Regulation 12(1)(b)

182 Whereas regulation 12(1)(a) requires means to be provided whereby the supply of electrical energy can be switched off, 12(1)(b) requires that there will be available suitable means of ensuring that the supply will remain switched off and inadvertent reconnection prevented. This is isolation. This provision, in conjunction with safe working practices, will enable work to be carried out on electrical equipment without risk of it becoming live during the course of that work, eg if the work is to be done under the terms of regulation 13.

183 In some cases, the equipment used to perform the requirement under regulation 12(1)(a) may also serve to perform the requirement under 12(1)(b). The two functions of **switching off** and **isolation** are not the same, even though in some circumstances they are performed by the same action or by the same equipment.

Regulation 12(3)

184 Regulation 12(3) recognises the impracticability in some cases of switching off or isolating that equipment which is itself an integral part of a source of electrical energy, eg the terminals of batteries, battery cells, photovoltaic systems, large capacitors and the windings of generators. The regulation requires precautions to be taken in these circumstances so that danger is prevented so far as is reasonably practicable. See References and further reading for details of guidance on working practices.

'Where necessary to prevent danger'

185 The need for means to cut off the supply and effect isolation depends on factors such as likely danger in normal and abnormal conditions. This assessment may be influenced by environmental conditions and provisions to be made in case of emergencies, such as a fire in premises. It includes consideration of which electrical equipment could be a source of danger if such means were not provided and of the installation, commissioning, operational and maintenance requirements over the life of the equipment.

Suitable means for cutting off the supply

186 The suitable means for cutting off the supply (regulation 12(1)(a)) should:

(a) be capable of cutting off the supply under all likely conditions having regard to the equipment, its normal operation conditions, any abnormal operating or fault conditions, and the characteristics of the source(s) of electrical energy;
(b) be in a suitable location regarding the nature of the risks, the availability of people to operate the means and the speed at which operation may be necessary. Access to switches etc should be kept clear and unobstructed, free of tripping and slipping hazards etc;
(c) be clearly marked to show its relationship to the equipment which it controls, unless there could be no doubt that this would be obvious to any person who may need to operate it;
(d) only be common to several items of electrical equipment where it is appropriate for these to be energised and de-energised as a group.

Guidance 12

Suitable means of isolation

187 The suitable means of isolation of equipment (regulation 12(1)(b)) should:

(a) have the capability to positively establish an air gap or other effective dielectric which, together with adequate creepage and clearance distances, will ensure that there is no likely way in which the isolation gap can fail electrically;

(b) include, where necessary, means directed at preventing unauthorised interference with or improper operation of the equipment, eg means of locking off;

(c) be located so the accessibility and ease with which it may be employed is appropriate for the application. The time and effort that must be expended to effect isolation should be reasonable, depending on the nature of the equipment and the circumstances under which isolation may be required. For example, a very remote means of isolation may be acceptable if isolation is only needed infrequently and any additional time taken to effect isolation does not result in danger;

(d) be clearly marked to show which equipment it relates to, unless there could be no doubt that this would be obvious to any person who may need to operate it;

(e) only be common to several items of electrical equipment where it is appropriate for these to be isolated as a group.

Selection of isolator switches

188 Isolator switches (or disconnectors) will often be used as the means of disconnection and securing separation from the supply. When selecting appropriate equipment, consider:

(a) the isolating distances between contacts or other means of isolation which should meet an appropriate standard or be otherwise equally effective;

(b) the position of the contacts or other means of isolation, which should either be externally visible or clearly and reliably indicated. An indication of the isolated position, other than by direct observation of the isolating gap, should occur when the specified isolating distance has been achieved in each pole;

(c) provision to enable the prevention of unauthorised, improper or unintentional energisation, eg locking-off facilities.

189 For further information on the selection of isolators/disconnectors, refer to appropriate standards (see Further reading).

Regulation 13 Precautions for work on equipment made dead

Regulation 13

*Adequate precautions shall be taken to prevent **electrical equipment**, which has been made dead in order to prevent **danger** while work is carried out on or near that equipment, from becoming electrically charged during that work if **danger** may thereby arise.*

Guidance 13

190 The defence (regulation 29) is available in any proceedings for an offence under this regulation.

191 The regulation may apply during any work, be it electrical or non-electrical. The regulation requires adequate precautions to be taken to prevent the electrical

Guidance 13

equipment that has been made dead from becoming electrically charged, from whatever source, if this charging would give rise to danger. 'Charged' is discussed under regulation 2.

192 The regulation uses the term 'electrical equipment', which is defined by regulation 2 and explained in paragraphs 20–21.

The precautions

193 The precautions must be effective in preventing the electrical equipment from becoming charged in any way which would give rise to danger.

194 In the first place, the procedures for making the equipment dead will probably involve use of the means required by regulation 12(1)(a) for cutting off the supply of electrical energy. Isolation of the electrical equipment will be necessary and the means required by regulation 12(1)(b) will facilitate this. Ideally, a means of locking off an isolator can be used. Where such facilities are not available, the removal of fuses or links and their being held in safe keeping can provide a secure arrangement if proper control procedures are used.

195 These precautions will prevent the equipment from becoming charged by connection to its own or normal sources of electrical energy but may not, alone, be sufficient to prevent charging. The presence of electrical energy as a result of electromagnetic induction, mutual capacitance or stored electrical energy may have to be guarded against, eg by applying earthing connections for the duration of the work (temporary earths). The precautions may need to include means of preventing further accumulation of electrical charge, following initial discharge, because latent energy may be stored in the system, eg in the dielectric of high-voltage cables or capacitors within equipment. For work on high-voltage power distribution circuits, isolation procedures should include the application of circuit main earths (primary earths) at points of isolation and additional earthing around the point of work.

196 Where work is to be done on or near conductors that have been isolated, the conductors must be proved dead at the point of work before the task starts. Where a test instrument or voltage indicator is used for this purpose, this should itself be proved, preferably immediately before and immediately after testing the conductor (see also regulation 4(3)).

197 The regulation does not prevent the application of a test voltage to equipment, provided that this does not give rise to danger.

Written procedures

198 The safety isolation procedures should be formalised in written instructions or house rules. Safety documentation, including 'permits-to-work', may form part of the written procedures and their use is considered essential to ensuring a safe system of work where this involves work on the conductors or equipment of high-voltage power distribution systems (typically where the working voltage exceeds 1000 V) or where the system is very complex. Properly formulated and regulated 'permit-to-work' procedures focus the minds of those issuing and of those receiving the permits, both on the manner in which the work is to be done and on how the equipment has been made safe. Further advice on these procedures and precautions is in the guidance listed in References and further reading.

Guidance 13

Decommissioned equipment

199 Before electrical equipment is decommissioned, dismantled or abandoned for any reason, it must be disconnected from all sources of supply and effective steps taken to ensure that it is dead and cannot inadvertently become re-energised or dangerously charged. It may be necessary to securely mark or otherwise suitably label equipment, circuits, switches etc to guard against inadvertent re-energisation. (See also the requirement for identifying circuits under regulation 12(1).)

Regulation 14 Work on or near live conductors

Regulation 14

*No person shall be engaged in any work activity on or so near any live **conductor** (other than one suitably covered with insulating material so as to prevent **danger**) that **danger** may arise unless –*

(a) *it is unreasonable in **all** the circumstances for it to be dead; and*
(b) *it is reasonable in all the circumstances for him to be at work on or near it while it is live; and*
(c) *suitable precautions (including where necessary the provision of suitable protective equipment) are taken to prevent **injury**.*

Guidance 14

200 The defence (regulation 29) is available in any proceedings for an offence under this regulation.

201 Regulation 14 addresses the situation where, either permanently or temporarily, danger from conductors is not prevented by the precautions specified in regulation 7(a).

202 The regulation is concerned only with those situations where people are at work on or near live electrical conductors which may foreseeably give rise to danger. Such work is permitted only if conditions (a), (b) and (c) are satisfied. 'Work' is not confined to electrical work but includes any work activity, eg electrical testing.

The need for the conductor to be live

203 If danger may otherwise arise it is always preferable that work on or near electrical equipment should be carried out when that equipment is dead (see regulation 13 and guidance). Regulation 14 recognises that there are circumstances, however, in which it is unreasonable, having regard to all relevant factors, for the equipment to be dead while work proceeds. An example of this might be undertaking maintenance, checks or repairs on a busy section of electric railway track where it would be disproportionately disruptive and costly for the live conductors to be isolated for the period of the work. Other examples are in the electrical supply industry, particularly live cable jointing, and in much of the work done on telephone network connections.

204 When ordering, purchasing and installing plant, consider the manner of operation, maintenance and repair of the electrical equipment which will be necessary during its life.

205 The design of electrical equipment and of the installation should eliminate the need for live work which puts people at risk of injury. This can often be done by careful thought at the design stage of installations, for example by the provision of alternative power infeeds; properly laid out distribution systems to allow parts to be isolated for work to proceed; and by designing equipment housings etc which

Guidance 14

result in segregation of parts to be worked on and protect people from other parts which may be live.

206 Equipment which combines power and control circuitry should be arranged so that the power circuits are physically separate and segregated from logic and control circuits, or so placed, recessed or otherwise arranged that the risk of accidental contact is eliminated. Diagnostic work on the low-power/voltage circuits may then proceed with less risk to personnel. Where regular measurements of, say, voltage, current etc are to be made, consider appropriate test and measuring equipment, eg voltmeters, ammeters etc, or test points being built into the equipment.

207 Live work includes live testing, for example the use of a potential indicator on mains power and control logic circuits (but see paragraph 218).

208 The factors which should be considered in deciding whether it was justifiable for work to proceed with the conductors live should include the following:

(a) It is not practicable to carry out the work with the conductors dead, eg where for the purposes of testing it is necessary for the conductors to be live.
(b) To make the conductors dead will create other hazards, such as to other users of the system, or for continuously operating process plants etc.
(c) The need to comply with other statutory requirements.
(d) The level of risk involved in working live and the effectiveness of the precautions available set against the economic need to perform that work.

The need to be near uninsulated live conductors

209 People at work are permitted to be near live conductors only if this is reasonable in all the circumstances. If, for example, it would be reasonable for the work to be carried out at a safe distance from the conductors then it would be prohibited for that work to be done near the conductors.

210 People whose presence near the live conductors is not necessary should not be so near the conductors that they are at risk of injury. However, there may be occasions when people who do not normally need to be in the vicinity of live conductors are required to be present, eg those assisting or recovering a casualty after an incident. Appropriate measures must be put in place to protect them.

The need to take precautions to prevent injury

211 The precautions necessary to comply with regulation 14(c) should be commensurate with the risk.

212 The system of work must:

(a) allow only people who are competent to do so to work on or near exposed, live conductors (competence for these and other purposes is further dealt with at regulation 16); *and*
(b) indicate within what limits the work is to be attempted; *and*
(c) indicate what levels of competence apply to each category of such work; *and*
(d) incorporate procedures under which the person attempting the work will report back if the limits specified in the system are likely to be exceeded.

Guidance 14

213 This usually requires detailed planning before the work is started.

214 Suitable precautions should include, as appropriate:

(a) the use of people who are properly trained and competent to work on live equipment safely (see also regulation 16);
(b) the provision of adequate information to the person carrying out the work about the live conductors involved, the associated electrical system and the foreseeable risks (NB: when excavation work is ongoing the insulation of a cable is not an effective measure against damage to the cable or penetration of it by a mechanical tool);
(c) the use of suitable tools, including insulated tools, equipment and protective clothing (see also regulation 4(4));
(d) the use of suitable insulated barriers or screens (see also regulation 4(4));
(e) the use of suitable instruments and test probes;
(f) accompaniment by another person or people if the presence of such person or people could contribute significantly to ensuring that injury is prevented;
(g) the restriction of routine live test work (eg product testing) to specific areas and the use of special precautions within those areas, such as isolated power supplies, non-conducting locations etc;
(h) effective control of any area where there is danger from live conductors.

Accompaniment

215 A dutyholder's judgement as to whether someone carrying out work subject to regulation 14 should be accompanied, should be based on considerations of how injury is to be prevented. If an accompanying person can substantially contribute towards the implementation of safe working practice, then they should be present. They should be trained to recognise danger and, if necessary, to render assistance in the event of an emergency.

216 Some examples of electrical work where it is likely that the person carrying out the work should be accompanied are:

(a) electrical work involving manipulation of live, uninsulated power conductors at, say, 230 V using insulated tools;
(b) other work on or near bare live conductors where someone working on their own would not be capable of undertaking the work safely without assistance in, for example, keeping other people from the work area.

Control of the area

217 Where there is danger from live conductors ensure that those who are not competent to prevent the occurrence of injury, and those whose presence is unnecessary, are not permitted into the area. If the person undertaking the work is continuously present while danger exists from the live conductors, and the area is small enough to be under their constant supervision and control, then further precautions to control access may not be necessary. If, however, the area is too large or they are not continuously present, then effective control should be secured by other means, such as the provision of lockable enclosures or barriers and warning notices indicating the presence of live conductors. (The above examples are given without prejudice to the requirements of regulation 14, the criteria of which must be followed in each case before live work is undertaken.)

Testing

218 Regulation 14 will often apply to electrical testing. Testing to establish whether electrical conductors are live or dead should always be done on the assumption that they may be live and, therefore, it should be assumed that this regulation is applicable until such time as the conductors have been proved dead.

Guidance 14

219 When testing for confirmation of a 'dead' circuit, the test instrument or voltage indicator used for this purpose must itself be proved, preferably immediately before and immediately after testing the conductors.

220 Although live testing may be justifiable it does not follow that there will necessarily be justification for subsequent repair work to be carried out live.

Suitable protective equipment

221 Protective equipment suitable for the work activity should only be used as a last resort, ie when all other ways to eliminate or reduce risks have been considered.

222 Examples of equipment that can protect someone from the effects of electricity are:

(a) suitable clothing, including insulating helmets, goggles and gloves;
(b) insulating materials used as fixed or temporary screening to prevent:
 (i) electric shock;
 (ii) short circuit between live conductors;
 (iii) short circuit between live conductors and earth;
(c) insulating mats and stands to prevent electric shock current via the feet;
(d) insulated tools;
(e) insulated test probes.

223 There must be procedures for the periodic examination and, where necessary, testing and replacement of this protective equipment. See also the requirements of regulation 4(4), and References and further reading for details of guidance on working procedures, standards etc.

Emergency resuscitation and first aid

224 You should consider placing notices or placards giving details of emergency resuscitation procedures in the event of electric shock at those locations where people may be at greater risk of electric shock. Such places might include electrical test areas, substations and laboratories but, for resuscitation techniques to be effective, those required to exercise them must receive proper training and regular practice. The Health and Safety (First-Aid) Regulations 1981[11] make various requirements for the provision of suitably trained first-aiders at places of work.

Work near underground cables

225 Serious injuries have occurred during excavation and other work near underground power cables. This work comes within the scope of regulation 14 if there is a risk of injury from these cables.

226 Underground power cables present a risk of serious or fatal injury during excavation or similar work, particularly to people using hand tools (eg picks, concrete breakers etc). Precautions should include:

(a) mapping, recording and marking on site of cable runs;
(b) use of cable-locating devices;
(c) safe digging practices.

227 Well-established advice on working near underground cables is given in *Avoiding danger from underground services*.[12]

Work near overhead power lines

228 Every year workers are killed or injured while working near to overhead power lines. They may be readily accessible to people working on elevated platforms,

Guidance 14

scaffolding or roofs. People working with tall vehicles such as cranes, tipper lorries or farm machinery, or handling metal ladders, pipes or other long articles may also be at risk from a flashover or contact with overhead power lines.

229 Well-established advice on work near overhead power lines is given in HSE guidance notes *Avoiding danger from overhead power lines*[13] and *Working safely near overhead electricity power lines*.[14]

Regulation 15 Working space, access and lighting

Regulation 15

*For the purposes of enabling **injury** to be prevented, adequate working space, adequate means of access, and adequate lighting shall be provided at all **electrical equipment** on which or near which work is being done in circumstances which may give rise to **danger**.*

Guidance 15

230 The defence (regulation 29) is available in any proceedings for an offence under this regulation.

231 The purpose of the regulation is to ensure that sufficient space, access and adequate illumination are provided while people are working on, at or near electrical equipment so that they may work safely. The requirement is not restricted to those circumstances where live conductors are exposed, but applies where any work is being done in circumstances which may give rise to danger. The regulation does not require such space, access or illumination to be provided at times other than when work is being done. (But see guidance under regulation 12(1)(a) (paragraph 186(b)) in respect of safe access to means of cutting off the supply.)

Working space

232 Where there are dangerous exposed live conductors within reach, the working space dimensions should be adequate:

(a) to allow people to pull back away from the conductors without hazard;
(b) to allow people to pass one another with ease and without hazard.

233 Among the legal provisions revoked when these Regulations came into force were the Electricity (Factories Act) Special Regulations 1908 and 1944. Regulation 17 of those Regulations specified minimum width and height dimensions of switchboard passageways where there were bare conductors exposed or arranged to be exposed when live so that they may be touched. That regulation and the relevant definitions used are reproduced in Appendix 1. The dimensions specified were arrived at after much consideration of the circumstances in a Public Inquiry when those Regulations were being drafted. However, those dimensions can still provide guidance for an appropriate level of safety in many circumstances and where the voltages do not significantly exceed 3000 V. This does not condone the use of equipment having normally bare and exposed conductors if a safe alternative can reasonably be adopted.

Lighting

234 Natural light is preferable to artificial light, but where artificial light is necessary it is preferable that this be from a permanent and properly designed installation – in indoor switchrooms etc. However, there will always be exceptions and special circumstances where these principles cannot be achieved, eg where handlamps or

Guidance 15

torches may be the sole or most important means of lighting. Whatever level of lighting is used, it must be adequate to enable injury to be prevented. HSE guidance *Lighting at work*[15] refers.

Regulation 16 Persons to be competent to prevent danger and injury

Regulation 16

*No person shall be engaged in any work activity where technical knowledge or experience is necessary to prevent **danger** or, where appropriate, **injury**, unless he possesses such knowledge or experience, or is under such degree of supervision as may be appropriate having regard to the nature of the work.*

Guidance 16

235 The defence (regulation 29) is available in any proceedings for an offence under this regulation.

236 The object of the regulation is to ensure that people are not placed at risk due to their own lack of competence in dealing with electrical equipment, or that of others.

'... prevent danger or, where appropriate, injury ...'

237 This regulation uses both terms: 'injury' and 'danger'. The regulation therefore applies to all work associated with electrical equipment where danger may arise and whether or not danger (or the risk of injury) is actually present during the work. It will include situations where the elimination of the risk of injury, ie the prevention of danger, for the duration of the work is under the control of someone who must therefore possess sufficient technical knowledge or experience – or be so supervised etc – to be capable of ensuring that danger is prevented. For example, where a person is to isolate some electrical equipment before undertaking work on the equipment, they will require sufficient technical knowledge or experience to prevent danger during the isolation. There will be no danger from the equipment during the work, provided that the isolation has been carried out properly: danger will have been prevented. However, the person doing the work must have sufficient technical knowledge or experience so as to prevent danger during that work, eg by knowing not to work on adjacent 'live' circuits.

238 The regulation also covers those circumstances where danger is present, ie where there is a risk of injury, as for example where work is being done on live or charged equipment using special techniques and under the terms of regulation 14. In these circumstances, people must possess sufficient technical knowledge or experience – or be so supervised etc – to be capable of ensuring that injury is prevented.

Technical knowledge or experience

239 The scope of 'technical knowledge or experience' should include:

(a) adequate knowledge of electricity;
(b) adequate experience of the electrical work being carried out;
(c) adequate understanding of the system to be worked on and practical experience of that class of system;
(d) understanding of the hazards which may arise during the work and the precautions which need to be taken;
(e) the ability to recognise at all times whether it is safe for work to continue.

Guidance 16

Allocation of responsibilities

240 Employees must be trained and instructed to ensure that they understand the safety procedures which are relevant to their work and must work in accordance with any instructions or rules laid down by their employer and directed at ensuring safety.

Supervision

241 The regulation recognises that, in many circumstances, people will require some degree of supervision where they do not have sufficient technical knowledge or experience to ensure that they can undertake the work safely. Dutyholders, when allocating to supervisors responsibilities for supervision, should clearly state to the supervisor exactly what their responsibilities are and consider stating these responsibilities in writing. Where the risks involved are low, verbal instructions are likely to be adequate but as the risk or complexity increase there comes a point where the need for written procedures becomes important if instructions are to be understood and supervised more rigorously. In this context, supervision does not necessarily require continual attendance at the work site, but the degree of supervision and the manner in which it is exercised is for the dutyholders to arrange to ensure that danger, or as the case may be, injury, is prevented.

242 You should also refer to appropriate guidance, such as that in national, international, reputable foreign and harmonised or industry standards and codes of practice, or to HSE guidance, or seek expert advice.

Regulations 17–28

243 Regulations 17 to 28 and Schedule 1 inclusive were revoked by the Mines Regulations 2014.

Regulation 29 Defence

Regulation 29

In any proceedings for an offence consisting of a contravention of [regulations 4(4), 5, 8, 9, 10, 11, 12, 13, 14, 15 or 16,]* it shall be a defence for any person to prove that he took all reasonable steps and exercised all due diligence to avoid the commission of that offence.

* Amended by the Mines Regulations 2014, SI 2014/3248.

Guidance 29

244 Regulation 29 applies only in criminal proceedings. It provides a defence for a dutyholder who can establish that they took all reasonable steps and exercised all due diligence to avoid committing an offence under regulations 4(4), 5, 8, 9, 10, 11, 12, 13, 14, 15 or 16.

Regulation 30 Exemption certificates

Regulation 30

(1) Subject to paragraph (2), the Health and Safety Executive may, by a certificate in writing, exempt –

(a) any person;
(b) any premises;
(c) any electrical equipment;
(d) any electrical system;
(e) any electrical process;
(f) any activity,

or any class of the above, from any requirement or prohibition imposed by these Regulations and any such exemption may be granted subject to conditions and to a limit of time and may be revoked by a certificate in writing at any time.

(2) The Executive shall not grant any such exemption unless, having regard to the circumstances of the case, and in particular to –

(a) the conditions, if any, which it proposes to attach to the exemption; and
(b) any other requirements imposed by or under any enactment which apply to the case,

it is satisfied that the health and safety of persons who are likely to be affected by the exemption will not be prejudiced in consequence of it.

Guidance 30

245 HSE is given power to issue general or special exemptions and to impose conditions and time limits on them. It is a standard power given to allow the variation of legal duties where, in circumstances unforeseen by those drafting the legislation, they are in practice unnecessary or inappropriate. Exemptions would be granted only in very exceptional circumstances.

Regulation 31 Extension outside Great Britain

Regulation 31

These Regulations shall apply –

(a) in Great Britain; and
(b) outside Great Britain as sections 1 to 59 and 80 to 82 of the Health and Safety at Work etc Act 1974 apply by virtue of the provisions of the Health and Safety at Work etc Act 1974 (Application outside Great Britain) Order 1995.

Guidance 31

246 Regulation 31 was modified by the Offshore Electricity and Noise Regulations 1997. Although the regulation refers to the Health and Safety at Work etc Act 1974 (Application outside Great Britain) Order 1995, this order has been revoked. The regulation should be read as referring to the Health and Safety at Work etc Act 1974 (Application outside Great Britain) Order 2013 (SI 2013/240).[16]

247 The Electricity at Work Regulations apply to all work activities on offshore installations, wells, pipelines and pipelines works and to certain connected activities within the territorial waters of Great Britain, or in the designated areas of the UK Continental Shelf. They also apply to certain other activities within territorial waters, including the construction and operation of wind farms.

Regulation 32 Disapplication of duties

The duties imposed by these Regulations shall not extend to –

(a) the master or crew of a sea-going ship or to the employer of such persons, in relation to the normal ship-board activities of a ship's crew under the direction of the master; or

(b) any person, in relation to any aircraft or hovercraft which is moving under its own power.

Sea-going ships

248 Sea-going ships are subject to other electrical safety legislation which gives protection to people on board. Regulation 32 disapplies the Electricity at Work Regulations from these ships as far as the normal ship-board activities of a ship's crew under the direction of the master is concerned. It does not disapply them in respect of other work activities however, eg where a shore-based electrical contractor goes on board to carry out electrical work on the ship. That person's activities will be subject to the Regulations within the general applicability of the Regulations. The Regulations will apply outside Great Britain only as provided for under regulation 31.

Aircraft and hovercraft

249 The Regulations may apply only while an aircraft or hovercraft is not moving under its own power.

Vehicles

250 The Regulations may apply to electrical equipment on vehicles if this equipment may give rise to danger.

Regulation 33 Revocations and modifications

(1) The instruments specified in column 1 of Part I of Schedule 2 are revoked to the extent specified in the corresponding entry in column 3 of that Part.

(2) The enactments and instruments specified in Part II of Schedule 2 shall be modified to the extent specified in that Part.

(3) In the Mines and Quarries Act 1954, the Mines and Quarries (Tips) Act 1969 and the Mines Management Act 1971, and in regulations made under any of those Acts, or in health and safety regulations, any reference to any of those Acts shall be treated as including a reference to these Regulations.

251 The Regulations replace or modify a number of statutory provisions in accordance with the intention of the HSW Act section 1(2).

252 Systems and equipment which were subject to provisions which have been revoked are now subject to these Regulations.

Appendix 1

Working space and access: Historical comment on revoked legislation (see regulation 15)

Among the legal provisions revoked upon the coming into force of the Electricity at Work Regulations 1989 were the Electricity (Factories Act) Special Regulations 1908 and 1944. Regulation 17 of those Regulations specified minimum width and height dimensions of 'switchboard passage-ways' if there were 'bare conductors' exposed or arranged to be exposed when 'live' so that they may be touched. These related to what are commonly known as 'open type' switchboards which had much exposed copper work, knife switches etc. That regulation (and the key definitions used at that time) are reproduced below for information. The dimensions which were specified by that regulation were arrived at after much consideration of the circumstances at the time. A compromise was struck between the objective of achieving the safety of those who had to work at and operate these 'open type' switchboards and the need to recognise the constraints imposed by the installations existing and the nature of the technology in 1908. Even though the dimensions were a compromise, it was widely recognised that they were a good minimum standard which had been found necessary following a number of severe and fatal accidents in factories and power stations due to inadequate space or cluttered access in the vicinity of bare live conductors at these 'open type' switchboards. The dimensions chosen allowed workmen to operate or otherwise work upon the switchboard in reasonable safety and allowed, for example, people to pass one another in the switchboard passageway without being placed at unacceptable risk of touching live conductors.

Where the need does arise to work on or near live conductors, the principles of providing adequate working space and uncluttered access/egress, which were expressed in regulation 17 of the Electricity (Factories Act) Special Regulations 1908 and 1944, should be given proper consideration.

Regulation 17 (of 1908 Regulations)

At the working platform of every switchboard and in every *switchboard passage-way*, if there be *bare conductors* exposed or arranged to be exposed when *live* so that they may be touched, there shall be a clear and unobstructed passage of ample width and height, with a firm and even floor. Adequate means of access, free from danger, shall be provided for every *switchboard passage-way*.

The following provisions shall apply to all such *switchboard* working platforms and passage-ways constructed after January 1, 1909 unless the bare conductors, whether overhead or at the sides of the *passage-ways*, are otherwise adequately protected against *danger* by divisions or screens or other suitable means:

(a) Those constructed for *low pressure* and *medium pressure switchboards* shall have a clear height of not less than 7 ft and a clear width measured from *bare conductor* of not less than 3 ft.

(b) Those constructed for *high pressure* and *extra high pressure switchboards*, other than operating desks or panels working solely at *low pressure*, shall have a clear height of not less than 8 ft and a clear width measured from *bare conductor* of not less than 3 ft 6 in.

(c) Bare conductors shall not be exposed on both sides of the *switchboard passage-way* unless either (i) the clear width of the passage is in the case of *low pressure* and *medium pressure* not less than 4 ft 6 in and in the case of *high pressure* and *extra high pressure* not less than 8 ft in each case measured between bare conductors, or (ii) the *conductors* on one side are so guarded that they cannot be accidentally touched.

Key definitions used in the 1908 Regulations

Switchboard means the collection of switches or fuses, *conductors*, and other *apparatus* in connection therewith, used for the purpose of controlling the current or pressure in any *system* or part of a *system*.

Switchboard passage-way means any passage-way or compartment large enough for a person to enter, and used in connection with a *switchboard* when *live*.

Low pressure means a *pressure* in a *system* normally not exceeding 250 volts where the electrical energy is used.

Medium pressure means a *pressure* in a *system* normally above 250 volts, but not exceeding 650 volts, where the electrical energy is used.

High pressure means a *pressure* in a system normally above 650 volts, but not exceeding 3000 volts, where the electrical energy is used or supplied.

Extra-high pressure means a *pressure* in a *system* normally exceeding 3000 volts where the electrical energy is used or supplied.

References and further reading

References

1 *Electrical safety in mines* HSG278 HSE Books 2015 www.hse.gov.uk/pubns/books/hsg278.htm

2 *The Health and Safety at Work etc Act 1974 (c37)* The Stationery Office 1974 ISBN 978 0 10 543774 1

3 *Managing for health and safety* HSG65 (Third edition) HSE Books 2013 www.hse.gov.uk/pubns/books/hsg65.htm

4 BS 7671 *Requirements for electrical installations. IET Wiring Regulations. Seventeenth edition* British Standards Institution

5 *The Electricity Safety, Quality and Continuity Regulations 2002 (ESQCR)* SI 2002/2665 The Stationery Office www.legislation.gov.uk

6 *Guide to effects of current on human beings and livestock. Special aspects relating to human beings* PD 60479-2:1987 IEC www.iec.ch

7 *Safety of equipment electrically connected to a telecommunication network* 62151:2000 IEC www.iec.ch

8 *Electricity at work: Safe working practices* HSG85 (Third edition) HSE Books 2013 www.hse.gov.uk/pubns/books/hsg85.htm

9 *Dangerous substances and explosive atmospheres: Dangerous Substances and Explosive Atmospheres Regulations 2002. Approved Code of Practice and guidance* L138 (Second edition) HSE Books 2013 www.hse.gov.uk/pubns/books/l138.htm

10 *Maintaining portable electric equipment in low-risk environments* Leaflet INDG236(rev3) HSE Books 2013 www.hse.gov.uk/pubns/indg236.htm

11 *First aid at work: The Health and Safety (First-Aid) Regulations 1981. Guidance on Regulations* L74 (Third edition) HSE Books 2013 www.hse.gov.uk/pubns/books/l74.htm

12 *Avoiding danger from underground services* HSG47 (Third edition) HSE Books 2013 www.hse.gov.uk/pubns/books/hsg47.htm

13 *Avoiding danger from overhead power lines* General Guidance Note GS6 (Fourth edition) HSE 2013 www.hse.gov.uk/pubns/gs6.htm

14 *Working safely near overhead electricity power lines* AIS8(rev3) HSE Books 2012 www.hse.gov.uk/pubns/ais8.htm

15   *Lighting at work* HSG38 (Second edition) HSE Books 1998 www.hse.gov.uk/pubns/books/hsg38.htm

16   *The Health and Safety at Work etc Act 1974 (Application outside Great Britain) Order* 2013 SI 2013/240 The Stationery Office

Further reading

Guidance

Electrical safety and you: A brief guide Leaflet INDG231(rev1) HSE Books 2012 www.hse.gov.uk/pubns/indg231.htm

Electrical risks from steam/water pressure cleaners Plant and Machinery Guidance Note PM29 (Second edition) HSE Books 1995 www.hse.gov.uk/pubns/pm29.htm

Selection and use of electric handlamps Plant and Machinery Guidance Note PM38 (web only) HSE Books 2007 www.hse.gov.uk/pubns/pm38.htm

Electrical test equipment for use by electricians General Guidance Note GS38 (Fourth edition) HSE Books 2015 www.hse.gov.uk/pubns/gs38.htm

Electrical safety at places of entertainment General Guidance Note GS50 (Third edition) HSE Books 2014 www.hse.gov.uk/pubns/gs50.htm

Keeping electrical switchgear safe HSG230 HSE Books 2015 www.hse.gov.uk/pubns/books/hsg230.htm

Listed below are standards, codes of practice and other publications which contain guidance relevant to these Regulations and electrical safety – published by bodies other than HSE. Most of these documents are the product of technical committees on which HSE has been represented. This does not mean, however, that the documents are concerned solely with safety and users should bear in mind the scope of the safety content of these documents and the fact that they have largely been arrived at through a process of consensus.

International Electrotechnical Commission (IEC) publications

Guide to effects of current on human beings and livestock. Part 1: General aspects TS 60479-1:2005 IEC www.iec.ch

Safety of equipment electrically connected to a telecommunication network 62151:2000 IEC www.iec.ch

Extra-low voltage (ELV) – Limit values 61201:ed 2.0 2007–08 IEC www.iec.ch

Effects of current on human beings and livestock. General aspects DD IEC/TS 60479-1 www.iec.ch

Guide to effects of current on human beings and livestock. Special aspects relating to human beings PD 6519-2:1998 IEC 60479-2 www.iec.ch

Safety of equipment electrically connected to a telecommunication network IEC 62151 www.iec.ch

Extra-low voltage (ELV) – Limit values IEC 61201 ed 2.0 www.iec.ch

British Standards

The following British Standards are relevant to different aspects of electrical safety. The up-to-date version of each standard is available from BSI on http://shop.bsigroup.com (see Further information).

BS 4363 *Specification for distribution assemblies for reduced low voltage electricity supplies for construction and building sites*

BS 4444 *Guide to electrical earth monitoring and protective conductor proving*

BS 6423 *Code of practice for maintenance of low-voltage electrical switchgear and controlgear*

BS 6626 *Maintenance of electrical switchgear and control gear for voltages above 1 kV and up to and including 36 kV. Code of practice*

BS 6867 *Maintenance of electrical switchgear for voltages above 36 kV. Code of practice*

BS 7375 *Distribution of electricity on construction and building sites. Code of practice*

BS 7430 *Code of practice for protective earthing of electrical installations*

BS 7671 *Requirements for Electrical Installations. IET Wiring Regulations. Seventeenth edition*

BS 7909 *Code of practice for temporary electrical systems for entertainment and related purposes*

BS EN 50050 *Electrostatic hand-held spraying equipment. Safety requirements (50050 series, see Parts 1, 2 & 3)*

BS EN 60034 *Rotating electrical machines. Degrees of protection provided by the internal design of rotating electrical machines (IP Code). Classification*

BS EN 60079 *Explosive atmospheres (60079 series, see Parts 0, 14, 17, 31)*

BS EN 60204-1 *Safety of machinery. Electrical equipment of machines. General requirements*

BS EN 60529 *Degrees of protection provided by enclosures (IP Code)*

BS EN 60903 *Live working. Gloves of insulating material*

BS EN 60947 *Low-voltage switchgear and controlgear (60947 series, see Parts 1, 2)*

BS EN 61111 *Live working. Electrical insulating matting*

BS EN 61439-1 *Low-voltage switchgear and controlgear assemblies. General rules*

BS EN 61439-2 *High-voltage switchgear and controlgear. Alternating current circuit breakers*

BS EN 62271 *High-voltage switchgear and controlgear (62771 series, see Parts 100, 102)*

BS EN 62305-1 *Protection against lightning. General principles*

BS EN 81-20 *Safety rules for the construction and installation of lifts. Lifts for the transportation of persons and goods. Passenger and goods passenger lifts*

PD CLC/TR 50404 *Electrostatics. Code of practice for the avoidance of hazards due to static electricity*

Useful web links
www.hse.gov.uk/construction/safetytopics/electricity.htm

International Electrotechnical Commission (IEC) www.iec.ch

Further information

For information about health and safety visit https://books.hse.gov.uk or http://www.hse.gov.uk. You can view HSE guidance online and order priced publications from the website. HSE priced publications are also available from bookshops.

To report inconsistencies or inaccuracies in this guidance email: commissioning@williamslea.com.

British Standards can be obtained in PDF or hard copy formats from BSI: http://shop.bsigroup.com or by contacting BSI Customer Services for hard copies only Tel: 0845 086 9001 email: cservices@bsigroup.com.

The Stationery Office publications are available from The Stationery Office, PO Box 29, Norwich NR3 1GN Tel: 0333 202 5070 Fax: 0333 202 5080 email: customer.services@tso.co.uk Website: www.tsoshop.co.uk. (They are also available from bookshops.)

Statutory Instruments can be viewed free of charge at www.legislation.gov.uk where you can also search for changes to legislation.